# NEUROPLASTICITY

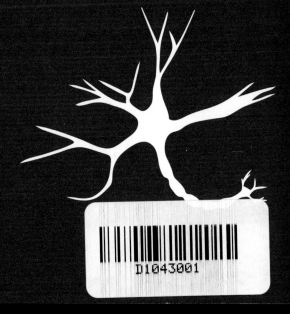

**The MIT Press Essential Knowledge Series**

# CONTENTS

# SERIES FOREWORD

The MIT Press Essential Knowledge series offers accessible, concise, beautifully produced pocket-size books on topics of current interest. Written by leading thinkers, the books in this series deliver expert overviews of subjects that range from the cultural and the historical to the scientific and the technical.

In today's era of instant information gratification, we have ready access to opinions, rationalizations, and superficial descriptions. Much harder to come by is the foundational knowledge that informs a principled understanding of the world. Essential Knowledge books fill that need. Synthesizing specialized subject matter for nonspecialists and engaging critical topics through fundamentals, each of these compact volumes offers readers a point of access to complex ideas.

*Bruce Tidor*
*Professor of Biological Engineering and Computer Science*
*Massachusetts Institute of Technology*

In recent years, researchers have made major advances in their understanding of how the brain works. Neuroscience offers both promises and perils, so public interest in it has grown enormously, because of the potential and promises that it offers, and one neuroscientific concept in particular has captured the public imagination: neuroplasticity, or change in the nervous system.

Sixty years ago, the idea that nervous tissue can change was anathema to neuroscience. It was widely believed that the mature brain is a fixed structure and, therefore, that "you can't teach an old dog new tricks." This dogma has since been overturned by a huge body of research which shows not only that the brain can change, but also that it changes continuously throughout life, in one way or another, in response to everything we do and every experience we have.

Neuroplasticity is a catch-all term referring to the many different ways in which the nervous system can change. It is ill-defined by neuroscientists, who use it to describe a wide variety of phenomena. Among the general public, the concept is generally misunderstood, and misconceptions about what neuroplasticity is, and what it is capable of, are rife.

This book summarizes essential knowledge about neuroplasticity for the general reader, focusing on key experimental findings and the methods and technologies used to obtain them, and on how our thinking about the brain has evolved over time.

Each chapter of the book is devoted to research into one particular mode of neuroplasticity. After offering a historical perspective, it describes the changes that occur during brain development; the changes that occur in the brains of blind and deaf people; the mechanisms of synaptic plasticity that underlie learning; the formation of new cells in the adult brain; how different types of training, such as learning to play a musical instrument or speak a foreign language, alter the brain; how neuroplasticity goes awry in addiction and pain states; and the key brain changes associated with different stages of life. The concluding chapter summarizes key points, describes several newly discovered forms of plasticity, and explores some of the many questions that remain to be answered.

The mechanisms of neuroplasticity are so many, and so diverse, that one could not hope to cover the whole subject in such a small book. Nevertheless, this volume should give the reader a good overview of what we know about neuroplasticity, as well as a solid understanding of some key neuroscientific principles, knowledge of important historical developments in the field, and some insight into the scientific process.

# INTRODUCTION

If you query Google about "rewiring your brain," its auto-complete function will give you a list of the most popular search terms using that phrase. You can, according to the results of such a search, rewire your brain for love and for happiness, to become more successful at work, and even to find meaning in your life. Scrolling down the search results brings up yet more options: rewire your brain to think positively, cultivate self-confidence, sleep better, and avoid procrastination. If the Internet is to be believed, you can rewire your brain to improve just about any aspect of your behavior, and so the power to transform your life lies in your ability to consciously change that 1.4-kilogram lump of meat inside your head.

But what does "rewiring your brain" actually mean? It refers to the concept of neuroplasticity, a very loosely defined term that simply means some kind of change in the nervous system. Just 50 years ago, the idea that the adult brain can change in any way was heretical. Researchers accepted that the immature brain is malleable, but also believed that it gradually hardens, like clay poured into a mold, into a permanently fixed structure by the time childhood has ended. It was also believed that we are born with all the brain cells we will ever have, that the brain is incapable of regenerating itself, and, therefore, that any damage or injuries it sustains cannot be fixed.

In fact, nothing could be further from the truth. The adult brain is not only capable of changing, but it does so continuously throughout life, in response to everything we do and every experience we have. Nervous systems evolved to enable us to adapt to the environment and determine the best course of action in any given situation, based on what has been learned from past experiences. This is the case not just for humans, but for all organisms that have a nervous system. That is to say, nervous systems evolved to change, and so neuroplasticity is an intrinsic and fundamental property of all nervous systems.

The concept of neuroplasticity therefore pervades every branch of brain research, and neuroscientists take it for granted that any experiment they perform will induce some kind of change in the nervous system of the organism they are studying. Different researchers define neuroplasticity in different ways, depending on exactly which aspect of brain and behavior they are studying, and the term is so vague that it has become virtually meaningless

The adult brain is not only capable of changing, but it does so continuously throughout life, in response to everything we do and every experience we have.

when used alone and without further explanation of exactly what type of plastic changes are taking place. Nevertheless, the idea that we can willfully shape our brains to change ourselves is an attractive one, and so the concept has captured the public imagination.

Today, neuroplasticity is a buzzword in many different realms. "Rewire your brain" has become something of a mantra for motivational speakers and self-help gurus, and the concept is being evoked by educationalists and business managers in their attempts to enhance learning and improve leadership skills. Misconceptions abound, however, and in these contexts neuroplasticity is usually ill-defined and often misunderstood. Some believe it has miraculous healing powers, and others say they can harness it with products or New Age therapies; but such claims are often hugely exaggerated and sometimes completely unfounded.

## A Brief History of Neuroplasticity

Neuroplasticity is often portrayed as a revolutionary new discovery, but the concept has existed in one form or another for over 200 years. In the early 1780s, correspondences between the Swiss naturalist Charles Bonnet and the Italian anatomist Michele Vincenzo Malacarne discuss the possibility that mental exercise can lead to brain growth, and mention various ways to test the idea experimentally. Malacarne then did so, using pairs of dogs from the same litter and pairs of birds from the same clutch of eggs. He trained one animal from each pair extensively for several years then examined their brains, and claimed that the cerebellum was significantly larger in the trained animals than in the untrained ones.

Shortly afterward, the German physician Samuel Thomas von Sömmerring entertained the idea in an influential anatomy textbook published in 1791: "Does use and exertion of mental power gradually change the material structure of the brain," he wrote, "just as we see, for example, that much used muscles become stronger and that hard labor thickens the epidermis considerably? It is not improbable, although the scalpel cannot easily demonstrate this."

In the early nineteenth century, Johann Spurzheim, one of the founders of phrenology, suggested that development of the mental faculties and the brain structures associated with them could be stimulated by exercise and education. And Jean-Baptiste Lamarck, an opponent of Charles Darwin who argued that evolution occurs by the inheritance of acquired characteristics, believed that specialized brain regions develop through proper use of the related faculties.[1]

In the 1830s, the physiologist Theodore Schwann and the botanist Matthias Schleiden developed cell theory, which stated that cells are the basic structural units of all living things. The microscopes available at the time were not powerful enough to resolve the finer details of nervous tissue, however. It was still not clear whether cell theory also applied to the nervous system, and throughout the nineteenth century there was debate about the fine structure of the brain and spinal cord. Researchers were split into two camps: the neuronists, who believed that the nervous system must, like all other living things, be made of cells, and the reticularists, who argued that it is made up of a continuous sheet of tissue.

The debate was finally settled in the 1890s, thanks largely to the work of the Spanish neuroanatomist Santiago Ramón y Cajal. Exploiting advances in microscopy and new staining methods, Cajal examined and compared nervous tissue from different species, including humans, and, being an accomplished artist, documented his observations in beautiful drawings. Drawing on his own work, and that of various others, he amassed sufficient evidence to convince the scientific community that nervous tissue is made of cells called neurons, which form contacts with each other. In so doing, he established modern neuroscience as a discipline in its own right, and today is considered to be its founding father.[2]

Darwin speculated about neuroplasticity in *The Descent of Man*, published in 1874. "I have shown that the brains of domestic rabbits are considerably reduced in

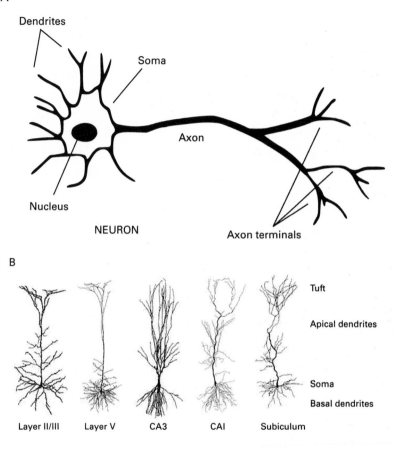

A

Dendrites

Soma

Axon

Nucleus

NEURON

Axon terminals

B

Tuft

Apical dendrites

Soma

Basal dendrites

Layer II/III    Layer V    CA3    CAI    Subiculum

**Figure 1** (A) Schematic diagram showing the main structures of a nerve cell (https://commons.wikimedia.org/wiki/Neuron#/media/File:Neuron_-_annotated.svg, CC BY-SA 3.0). (B) Pyramidal neurons from different layers and regions of the cerebral cortex, based on drawings by Cajal.

bulk, in comparison with those of the wild rabbit or hare," he wrote, "and this may be attributed to their having been closely confined during many generations, so that they have exerted their intellect, instincts, senses and voluntary movements but little."

But the term "plasticity" first appears in an 1890 textbook called *The Principles of Psychology* by William James. Here, James defines plasticity as "the possession of a structure weak enough to yield to an influence, but strong enough not to yield all at once," and explains habit formation in terms of strengthening of synapses and the formation of new connections: "If habits are due to the plasticity of materials to outward agents, we can immediately see to what outward influences, if to any, the brain-matter is plastic... and it is to the infinitely attenuated currents that pour in through [the sensory nerve-roots] that the hemispherical cortex shows itself to be so peculiarly susceptible. The currents, once in, must find a way out. In getting out they leave their traces in the paths which they take. The only thing they *can* do, in short, is to deepen old paths or to make new ones."

In 1894, Cajal suggested that plasticity occurs at the junctions between nerve cells and that mental exercise leads to growth of new nerve fiber branches. "The theory of free arborization of cellular branches capable of growing seems not only to be very probable but also most encouraging," he said in a lecture at the Royal Society in London. "A continuous pre-established network—a sort of system of telegraphic wires with no possibility for new stations or new lines—is something rigid and unmodifiable that clashes with our impression that the organ of thought is, within certain limits, malleable... especially during the developmental period... We could say that the cerebral cortex is like a garden planted with innumerable trees—the pyramidal cells—which, thanks to intelligent cultivation, can multiply their branches and sink their roots deeper, producing fruits and flowers of ever greater variety and quality."

Three years later, the British neurophysiologist Charles Sherrington named these junctions "synapses," from the Greek words *syn*, meaning "together," and *haptein*, meaning "to clasp," and stated that synapses are probably the sites at which learning takes place. He referred explicitly to synaptic strengthening: "Shut off from all opportunities of reproducing itself, the nerve cell directs its pent-up energy towards amplifying its connections with its fellows, in response to the events which stir it up."

Others challenged the notion that learning could induce new nerve fiber branches, pointing to evidence that there is far less variation in the size of the brain than in that of any other organ, and that brain volume appears to remain constant throughout much of life. Cajal preempted this objection by suggesting a "reciprocal diminution of the cell bodies or a shrinkage of other areas... whose function is not directly related to intelligence."

Less than 10 years later, however, Cajal appears to have changed his mind. "Once development was ended, the founts of growth of the axons and dendrites dried up irrevocably," he wrote in his 1913 textbook, *Degeneration and Regeneration of Nervous System*. "In the adult centers, the nerve paths are something fixed, ended, and immutable. Everything may die, nothing may be regenerated." This view quickly became one of the central dogmas of neuroscience, and researchers came to the general consensus that the brain is not materially affected by learning, experience, or training.[3]

### A Revolution in Modern Neuroscience

This dogma persisted well into the mid-twentieth century. In the early 1960s, however, the physiologists David Hubel and Torsten Wiesel made a series of seminal discoveries about how sensory experiences affect the developing brain, and the neuroscientist Paul Bach-y-Rita provided evidence that the adult human brain is not so fixed after all, using a "sensory substitution" device that enabled blind people to "see" with their sense of touch. Several other researchers reported that they had seen new cells being born in the brains of adult animals of various species, but were largely ignored, or ridiculed.

Then, in 1973, Tim Bliss and Terje Lømo reported the discovery of long-term potentiation (LTP), a physiological mechanism by which synapses could be strengthened for prolonged periods of time. This was another seminal discovery. Today, synaptic modification is widely regarded as the cellular basis of learning and memory, and as such, LTP is by far the most intensively studied and best understood mode of neuroplasticity. Since the initial discovery, researchers have accumulated a wealth of knowledge about the molecular mechanisms underlying LTP and related processes. Ironically, though, the work tells us very little about how learning and memory might be enhanced.

In the late 1990s, more direct evidence for neuroplasticity emerged, with the discovery of neural stem cells in the adult brain. This, more than anything, convinced the scientific community: the consensus shifted once again, and neuroplasticity was hailed as a revolutionary new discovery that overturned everything we thought we knew about the brain. Now, with more advanced technologies at their disposal, neuroscientists can visualize the brain in unprecedented detail and manipulate neuronal activity with great precision. These new methods have uncovered numerous other modes of neuroplasticity and also elucidated some of the underlying mechanisms.

Neuroplasticity can be seen in various forms at every level of nervous system organization, from the lowest levels of molecular activity and the structure and function of individual cells, through intermediate levels of discrete populations of neurons and widespread neuronal

Neuroplasticity can be seen in various forms at every level of nervous system organization, from the lowest levels of molecular activity to the highest level of brain-wide systems and behavior.

networks, to the highest level of brain-wide systems and behavior. Some occur continuously throughout life, others only at specific periods of life, and different types can be both induced separately and together.

Broadly speaking, there are two main types of neuroplasticity. Functional plasticity involves changes in some physiological aspect of nerve cell function, such as the frequency of nervous impulses or the probability of release of a chemical signal—both of which act to make synaptic connections stronger or weaker—or changes to the degree of synchronicity among populations of cells. Structural plasticity includes volumetric changes in discrete brain regions and the formation of new neural pathways, brought about either by the formation of new nerve fiber branches and synapses or by the growth and addition of new cells.

These different modes of plasticity occur over a wide range of timescales. Modification of synapses can occur on a timescale of milliseconds, synapses and dendrite branches are created or destroyed in the space of several hours, and new cells may be born or killed over periods of days. Other forms of neuroplasticity occur over even longer time frames—for example, brain maturation involves a protracted period of heightened plasticity that persists from late childhood into early adulthood, and losing one's sense of sight or hearing or sustaining brain damage induces gradual changes that occur in subsequent weeks, months, and years.

# SENSORY SUBSTITUTION

In the early 1800s, neurology began to flourish as researchers investigated the brain and formulated new theories about how its structure and function are related to behavior and mental functions.

During the first half of the century the field was dominated by phrenology, a pseudoscientific discipline that attempted to determine people's mental traits from skull measurements. This approach eventually fell into disrepute, giving way to another theory called the localization of cerebral function, according to which the brain is composed of discrete anatomical areas, each specialized to perform a specific function.

Subsequent work identified the sensory and motor regions of the brain, revealing not only that they are responsible for feeling and moving, respectively, but that these regions are always located in the same part of the brain. And so, when modern neuroscience was born, around the turn of the twentieth century, the idea that the cerebral cortex is composed of discrete regions specialized for language, touch, vision, and so on had already taken firm root.

With time, however, evidence began to emerge that the cortex is in fact highly plastic, and that the so-called modular organization of the brain is not set in stone. Much of this evidence comes from studies of blind and deaf people, whose brains have been completely deprived of a certain type of sensory input. Such work clearly shows that these cortical areas are not as specialized as we once thought—for example, the visual and auditory regions of the cortex can not only process information from other sense organs, but they can also contribute to non-sensory processes such as language.

## From Phrenology to the Localization of Cerebral Function

Phrenology was founded by the great anatomist Franz Joseph Gall, who stated that he first formulated his ideas at nine years of age. As a schoolchild, Gall had noticed that a classmate with a superior memory for words also had bulging eyes, and believed that the two characteristics appeared together in others. "Although I had no preliminary knowledge, I was seized with the idea that eyes thus formed were the mark of an excellent memory," he wrote.

With time, however, evidence began to emerge that the cortex is in fact highly plastic, and that this so-called modular organization is not set in stone.

"Later on... I said to myself; if memory shows itself by a physical characteristic, why not other features? And this gave me the first incentive for all my researches."

Gall began lecturing about phrenology in 1796, a year after graduating from medical school, and first published his theory in 1808. He came to believe that the region above the eyes was devoted to the "Faculty of Attending to and Distinguishing Words, Recollections of Words, or Verbal Memory." Later on, he documented the cases of two men who could not recall the names of relatives and friends as a result of sword injuries above the eye, which he took as confirmation of the early observations he had made at school.

He believed that "Destructiveness" resides above the ear, because this region was prominent in another school-child he knew, who was "fond of torturing animals," and in an apothecary who went on to become an executioner. He localized "Acquisitiveness" to another region slightly further back, because that region seemed to be disproportionately large in the pickpockets he had met; and "Ideality" to a region he believed to be prominent in statues of poets, writers, and other great thinkers, the area of the head they rubbed while writing.

Gall collected some 400 skulls throughout his career, including those of public intellectuals and psychopaths, and his theory was based almost exclusively on measurements he took from them. Overall, he claimed to have localized

27 mental faculties, and argued that 19 of them—including courage and the senses of space and color—could also be demonstrated in animals, whereas others—such as wisdom, passion, and a sense of satire—were unique to humans.

Though they faced criticism all along, the phrenologists remained influential up to the mid-nineteenth century. Their methods were eventually discredited as unscientific, however—Gall and his colleagues had "cherry-picked" their evidence, discarding any that was inconsistent with their theory—and so, by the 1870s the localization theory had become widely accepted, largely as the result of clinical investigations involving patients with brain damage.

In 1861, a French physician named Pierre Paul Broca described a handful of stroke patients who had been admitted to the hospital where he worked, all of whom had lost the ability to speak. Upon their death, Broca examined their brains, and noted that all of them were damaged in the same region of the left frontal lobe. Ten years later, the German pathologist Karl Wernicke described another group of stroke patients, who had lost the ability to understand spoken language due to damage affecting a region of the left temporal lobe.

Others found yet more evidence for the localization of cerebral function. Notably, the physiologists Gustav Fritz and Eduard Hitzig electrically stimulated and selectively destroyed parts of animals' brains; in doing so they localized the primary motor cortex to the precentral gyrus, and confirmed that this strip of brain tissue in each hemisphere controls movements of the opposite side of the body. But it was largely due to Broca's work that the cortical localization theory gained widespread acceptance.[1]

## The Brain Mappers

By the time modern neuroscience was born around the turn of the twentieth century, the idea that the cerebral cortex is composed of discrete anatomical regions with specialized functions was already firmly established. Even so, more evidence emerged in the early part of the twentieth century, and so the concept became further entrenched.

At around this time, a German neuroanatomist named Korbinian Brodmann began examining the microscopic structure of the human brain, and noticed that he could distinguish between different parts according to how the cells are organized in each. On this basis, Brodmann divided the cerebral cortex into 52 regions and assigned a number to each. Brodmann's system of neuroanatomical classification is still used to this day—Brodmann's areas 1, 2, and 3 make up the primary somatosensory cortex, which is located in the postcentral gyrus and receives touch information from the skin surface; Brodmann's area

4 is the primary motor cortex; and Brodmann's area 17 is the primary visual cortex.

In the 1920s, the Canadian neurosurgeon Wilder Penfield pioneered a technique to electrically stimulate the brains of conscious epilepsy patients, in order to determine the location of the abnormal brain tissue causing their seizures. Epilepsy can usually be treated effectively with anticonvulsant medication, but for the minority of patients who are unresponsive to drugs, surgery may be performed as a last resort, to remove the abnormal tissue and alleviate the debilitating seizures.

The brain is an extremely complex organ, and neurosurgery always runs the risk of causing collateral damage to areas involved in important functions such as language and movement. To avoid such damage, Penfield deliberately kept his patients conscious while he electrically stimulated the cortex, so that they could report their experiences back to him. When he stimulated the postcentral gyrus, for example, patients described feeling a touch sensation on some part of their body; stimulation of the precentral gyrus caused muscles in the corresponding part of the body to twitch; and stimulation of parts of the left frontal lobe interfered with the ability to speak. In this way he could delineate the boundaries of the abnormal tissue and remove it without inflicting damage on the surrounding tissue.

Penfield operated on approximately 400 patients, and in the process mapped the primary motor and somatosensory areas to the pre- and postcentral gyrus, respectively. He found that both strips of brain tissue are organized topographically, such that adjacent body parts are represented in adjacent regions of brain tissue (with a few minor exceptions); and that not all body parts are represented equally in the brain: the vast majority of the primary motor and somatosensory cortices are devoted to the face and hands, which are the most articulable and sensitive parts of the body.

Penfield summarized these important discoveries in *homunculus* ("little man") diagrams drawn up by his secretary. These drawings illustrated the organization of the primary motor and somatosensory cortices and the proportion of their tissues devoted to each body part, and were subsequently adapted into well-known three-dimensional models.[2]

### Sensory Substitution

Early evidence that this localization of brain function is not fixed came from studies performed in the late 1960s by Paul Bach-y-Rita, who built a device that enabled blind people to "see" with their sense of touch. The device consisted of a modified dentist's chair, fitted with 400 large vibrating pins arranged in a 20 by 20 array in the backrest, and connected to a large video camera that stood behind it on a large tripod.

Bach-y-Rita recruited a handful of blind people to test the apparatus, including a psychologist who had lost his sight at the age of four. To use it, the subjects simply sat in the chair and slowly moved the camera from side to side with a handle. As they did so, the image from the camera was converted into a pattern of vibrations on the array of pins in the backrest.

With extensive training, the subjects learned to use the touch sensations to interpret visual scenes accurately, beginning, after about an hour of training, with the ability to discriminate vertical, horizontal, diagonal, and curved lines, and then to recognize shapes. After more than about 10 hours of training, all of them could recognize common household objects, discern shadows and perspective, and even identify other people from their facial features.[3]

Bach-y-Rita argued that this ability was due to "cross-modal" mechanisms, whereby information that is normally conveyed by one sense, such as vision, is somehow transformed and conveyed by another, such as touch or sound. Since then, researchers have documented numerous examples of cross-modal plasticity, using modern neuroimaging techniques such as functional magnetic resonance imaging (fMRI) and transcranial magnetic stimulation (TMS).

Brain imaging studies reveal that the primary visual cortex is activated when blind people read braille, which requires fine motor control and touch discrimination to recognize the patterns of raised dots. This activation is associated with increased activity in downstream visual regions involved in shape recognition and with reduced activity in the somatosensory area, compared to sighted people. The same pattern is found not only in people who were born blind and those who lost their sight at an early age, but also in those who went blind later in life.

Interference with visual cortical activity, for example by the use of TMS, impedes touch perception in blind people but not in sighted controls, confirming that the activity in the visual cortex is indeed related to the processing of touch information, rather than merely coincidental.

Blind people can also learn to navigate by echolocation, by making clicking sounds with their tongue or tapping sounds with their feet, and using information in the returning echoes to perceive physical aspects of their surroundings. This requires huge amounts of training, but those who become adept at it can use echolocation to perform extremely complex actions that most of us could not imagine doing without sight, such as playing video games or riding a bike. And when blind people echolocate, the sound information is processed in visual rather than auditory parts of the brain.[4,5]

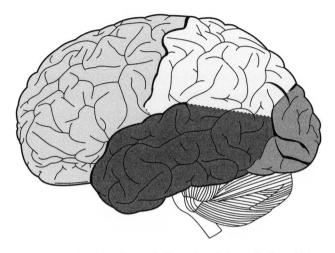

**Figure 2** The lobes of the brain. Clockwise, from left to right: frontal lobe, parietal lobe, occipital lobe, temporal lobe.

The visual system is often divided into two distinct pathways that run in parallel through the occipital lobe at the back of the brain—an upper stream that processes spatial information (the "where" pathway) and a lower one involved in object recognition (the "what" pathway). This organization seems to be preserved in the blind: when blind people learn to echolocate, the upper part of the visual cortex is activated when they are locating objects, and the lower part when they are identifying them.[6]

Thus, when deprived of the sensory inputs it normally receives, the visual cortex can switch roles and process other types of sensory information. Even more remarkably, it can adapt in such a way as to perform other, nonsensory functions, such as language. The same kind of brain scanning experiments show that this brain region is activated when blind people generate verbs, listen to spoken language, and perform verbal memory and high-level verbal processing tasks.

Blind people outperform sighted subjects on these tasks, and the extent of activation in their visual cortices is closely correlated to their performance in verbal memory tests. These studies also show that reading braille preferentially activates the front end of the visual cortex, whereas language activates the back region, and some find that the left visual cortex becomes more active than the right during language tasks, possibly because language centers are usually located in the left hemisphere. And just as interfering with visual cortical activity disrupts blind people's ability to process touch sensations and understand braille, so too does it impair their performance on verbal memory tasks.[7]

The brains of deaf people also show major plastic changes. In hearing people, sound information from the ears is processed by the auditory cortices in the temporal lobes. In people who are born deaf, however, these same brain areas are activated in response to visual stimuli. Deaf people also appear to have enhanced peripheral vision. This is linked to an increase in the overall area of the optic disc, where fibers of the optic nerve exit the eye on their way to the brain, and to thickening around its edges; it also suggests that the "where" stream of the visual pathway is stronger.

Neuroplasticity in deaf people is not confined to the visual and auditory systems. Using diffusion tensor imaging (DTI) to visualize brain connectivity, researchers have found that deafness is associated with major changes in long-range neural pathways, especially those between sensory areas of the cerebral cortex and a subcortical structure called the thalamus.

The thalamus has many important functions, particularly in relaying information from the sense organs to the appropriate cortical region, thus regulating the flow of information between different regions of the cortex. Deaf people exhibit changes in the microscopic structure of thalamus-cortex connections in every lobe of the brain, when compared to hearing people. Thus, deafness appears to induce brain-wide plastic changes that profoundly alter the way in which information flows through the brain.[8]

With advances in technology, sensory substitution devices have come a long way from Bach-y-Rita's cumbersome contraption. Rather than using them only as experimental tools, many research groups are now developing these devices as prosthetics that help blind and deaf people compensate for their sensory loss, and in June 2015 one such device was approved for use by the United States Food and Drug Administration (FDA). The BrainPort V100 is essentially a miniaturized version of Bach-y-Rita's apparatus, consisting of a video camera mounted onto a pair of sunglasses, and a 20 by 20 array of electrodes fitted onto a small, flat piece of plastic that is placed in the mouth. Computer software translates visual images from the camera, and transmits them to the electrodes, so that they are perceived as a pattern of tingling sensations on the tongue. In tests, about 70% of blind people learn to use the device to recognize objects after about a year of training.

## Cross-Modal Processing and Multisensory Integration

As studies of blindness and deafness show, the cerebral cortex has a remarkable capacity for plasticity, and the localization of brain function is not as strict as the neurologists of the nineteenth century believed it to be. Regions that are normally specialized to perform a specific function can switch roles and process other kinds of information, and the visual cortex in particular has been shown to be capable of performing a variety of nonvisual functions.

Under normal circumstances, the brain's sensory pathways are not entirely separate, but are interconnected and so can interact and influence each other in various ways. And while most primary sensory areas specialize in processing information from one particular sense organ, most of their downstream partners are so-called association

Regions that are normally specialized to perform a specific function can switch roles and process other kinds of information, and the visual cortex in particular has been shown to be capable of performing a variety of nonvisual functions.

areas, which combine various types of information in a process called multisensory integration.

Cross-modal processing and multisensory integration are important aspects of normal brain function, as the McGurk effect demonstrates. The McGurk effect is a powerful illusion that arises when there is a discrepancy between what we see and what we hear: the best example is a film clip of someone saying the letter *g*, dubbed with a voice saying the letter *b*, which is perceived as *d*. This consistent error clearly shows that vision and hearing interact and that the interaction aids our perception of speech.

Some researchers now argue that sensory substitution shares characteristics of, and is an artificial form of, a neurological condition called synesthesia, in which sensory information of one type gives rise to percepts in another sensory modality.[9] For example, the physicist Richard Feynman was a grapheme-color synesthete, for whom each letter of the alphabet elicited the sensation of a specific color, so that he saw colored letters when he looked at equations. The artist Wassily Kandinsky had another form of synesthesia. He experienced sound sensations in response to colors, and once said that he tried to create the visual equivalent of a Beethoven symphony in his paintings.

Once thought to be extremely rare, synesthesia is now believed to be relatively common, and may be experienced by one in every hundred people, or more. More than 40%

of synesthetes have a relative with the condition, indicating that genetics plays a big role. With training, however, non-synesthetes can learn to associate letters with colors or sounds, so that they evoke synesthetic experiences, and it is likely that this learning also occurs as a result of cross-modal plasticity.

Exactly how cross-modal plasticity arises is still unclear, but it probably involves a number of processes. During development, neural connections form somewhat haphazardly, and are then pruned back in response to sensory experiences that refine and fine-tune them (see chapter 3). Normally, most cross-modal connections are eliminated, but some remain in place for multisensory processing. Cross-modal plasticity may involve the "unmasking" of existing cross-modal connections and pathways that had been dormant, or the formation of entirely new ones, or both. Synesthesia may occur because of similar mechanisms, and the genes associated with it may play a role in preventing proper pruning of cross-modal pathways during brain development.

The question of how regions of the cerebral cortex become specialized to perform a particular function is particularly intriguing. Specialization is likely to occur as a result of both genetic and environmental factors. Cells in a given region are likely to activate specific combinations of genes that predispose them to perform a particular function, based on exactly where they are located and the connections they form. This blueprint can then be built on as sensory information sculpts the developing circuitry, or modified as necessary in the absence of one kind of information or another. Such a picture is supported by a 2014 study, which showed that deleting a single gene could respecify the identity of neurons in the primary somatosensory cortex of adult mice so that those cells processed information from other sensory modalities.[10]

# DEVELOPMENTAL PLASTICITY

The human brain is an organ of staggering complexity, containing 86 to 100 billion neurons, an even larger number of glial cells, and something on the order of a quadrillion exquisitely precise synaptic connections. Proper functioning of the nervous system depends upon all of these connections forming correctly; but how does such a complex organ develop? It has evolved a strategy of high redundancy—that is, the developing brain produces far more nerve cells than it actually needs but kills many of them off, and those that survive go on to form an overabundance of synaptic connections before pruning back the misguided or otherwise exuberant ones. These processes are governed by genetics but are highly dependent upon environment and experience during the early postnatal period, factors that fine-tune the neuronal circuits as they are being laid down.[1]

## Growth Factors and Cellular Suicide

In the late 1940s, a young Rita Levi-Montalcini joined the laboratory of the renowned embryologist Victor Hamburger and set about understanding the relationship between the developing nervous system and the organs and other tissues it innervates. In his own experiments, Hamburger had removed the developing limbs from chick embryos and noticed that primary sensory neurons, which extend fibers to the muscles in the limbs, did not survive in the absence of their "target" tissues. He concluded that nerve cells depend largely on their final destination to mature into a given type.

Levi-Montalcini speculated instead that removing target tissues caused the nerve cells to undergo some kind of degenerative process. Working together, they repeated Hamburger's experiments and confirmed his initial findings. Removing a limb bud caused the sensory neurons to die off and, conversely, grafting a supernumerary limb onto the embryo resulted in the survival of more cells. This led Levi-Montalcini to hypothesize that target tissues provide a feedback signal required for neuronal survival, and that the lack of that signal, or its low availability, causes cell death.[2]

Levi-Montalcini then continued trying to identify the signal and characterize its biological properties. Others had found that nerves rapidly grew into tumors that had been transplanted into chick embryos, leading Levi-Montalcini to the hypothesis that the transplanted tissue was secreting a diffusible factor that supported neuronal survival. Working with the biochemist Stanley Cohen, she added snake venom to sensory neurons growing in Petri dishes, and found that this actually produced more nerve fiber outgrowth than the tumors did.[3]

Cohen therefore suggested that they study mouse salivary glands, the mammalian equivalent of the snake venom gland. Fortuitously, they found the salivary glands to be a rich source of the feedback signal, so they succeeded in purifying the molecule and demonstrating that it was a small protein—which they called nerve growth factor (NGF). Cohen and Levi-Montalcini then went on to produce anti-NGF antibodies, and further demonstrated that the antibodies blocked the effect of the protein in newborn rodents and also on nerve cells growing in Petri dishes. These experiments showed conclusively that NGF is a diffusible protein that is secreted by certain tissues and promotes neuronal survival and differentiation.[4]

Levi-Montalcini's work provided direct evidence that extensive cell death occurs during neural development, and neatly explained how the nervous supply exactly matches the size of organs and other target tissues. According to the neurotrophic hypothesis, nerve cells are initially overproduced but then compete for a limited supply of target-derived NGF; those that receive the signal survive and undergo maturation, whereas those that do not wither and die.

NGF was the first growth factor to be identified. Its discovery and characterization was an important milestone in our understanding of neural development, and Levi-Montalcini and Cohen shared the 1986 Nobel Prize in Physiology or Medicine for their work.

Since then, the development of molecular biological techniques has led to the identification of dozens of other so-called neurotrophic factors, each of which promotes the survival of a specific population or populations of cells in the developing nervous system. The membrane receptor proteins that mediate their effects have been identified, too, and we are also beginning to understand some details about how they work: binding causes the growth factor–receptor protein complex to be internalized by the cell and then transported back to the nucleus, whereupon it can switch genetic programs on or off.[5]

It soon became clear that extensive cell death is a normal feature of neural development in all organisms. This process is called programmed cell death. It regulates the size of neuronal populations, the proper spacing and positioning of cells, and the emergence of shape and form, among other functions, and is therefore vital to proper development of the brain.

Cell death is under genetic control, and requires "executioner" genes that encode enzymes called caspases.

During development, the absence of neurotrophic signaling eventually switches these cell death genes on. Once the cellular suicide program has been activated, the caspase proteins begin to break the cell down from within: the cell's DNA and scaffold proteins are cut into fragments, causing chromosome condensation, cell shrinkage, and membrane blebbing, all of which give the dying cell a characteristic appearance. Finally, immune cells called macrophages engulf and clear away the cellular debris.[6]

## Synapse Formation

Immature neurons in the developing brain are highly promiscuous, forming many more synaptic connections than they need, before trimming back the exuberant, mismatched, and redundant ones.

Synapse formation (or *synaptogenesis*) is best understood at the neuromuscular junction, where the motor neuron nerve terminal comes into contact with skeletal muscle tissue. Cajal had recognized early on that these synapses are far more accessible and easier to study than those in the brain, which are much smaller and more densely packed. "Since the full grown forest turns out to be impenetrable and indefinable," he wrote in his autobiography, *Recollections of My Life*, "why not revert to the study of the young wood, in the nursery stage, as we might say?"

Immature neurons in the developing brain are highly promiscuous, forming many more synaptic connections than they need, before trimming back the exuberant, mismatched, and redundant ones.

At the neuromuscular junction, the motor neuron releases the neurotransmitter acetylcholine, which binds to receptors on the muscle fibers, causing them to contract. Initially, however, neither the nerve terminal nor the muscle is ready to perform this signaling process. The end of the developing nerve fiber takes the form of a growth cone—a dynamic structure covered with finger-like projections called filopodia—which detects chemical cues in the local environment to guide the growing tip of the nerve fiber to its proper destination, laying down new material as it proceeds. Likewise, the immature muscle mass has not yet split into individual muscle cells, and its acetylcholine receptor molecules are evenly distributed beneath the membrane.

Synapse formation and maturation are highly dependent upon interactions between the immature nerve and muscle. As the growth cone extends its tip to the muscle fiber, it releases a burst of acetylcholine, and this causes a redistribution of the acetylcholine receptors, which first cluster and then become immobilized at specific locations in the membrane. Innervation of the muscle by the nerve increases the conductance of the receptors already present in the muscle, and also elicits the synthesis of new receptor molecules, which are inserted into the muscle membrane.

Consequently, the muscle mass eventually splits into individual muscle fibers, each with a specialized receptor zone called an endplate. When the process is complete, there will be approximately 20,000 acetylcholine receptors per square micrometer of endplate, several thousand times the density of other regions of the muscle membrane.

At the earliest stages of development, neuronal growth cones split and send branches of the immature nerve fiber to more than one muscle fiber. But as development proceeds and the neurons mature, the number of synaptic connections gradually decreases. Spontaneous electrical activity stabilizes some of the connections, and experience strengthens them further. This process is also at least partly dependent upon the availability of growth factors in the muscle cells. Thus, nerve fiber branches that do not receive an adequate supply of growth factors are retracted, and synapses that are not strengthened by activity and experience are stripped away, such that every mature motor neuron innervates just one fiber.[7]

Synapses in the brain and spinal cord differ from the neuromuscular junction in several important ways. While the neuromuscular junction connects nerve to muscle, brain synapses connect neuronal elements to each other—a nerve terminal to a nerve cell body, an axon, or a dendrite. And whereas a mature motor neuron connects to a single muscle fiber, neurons in the brain are estimated to form an average of about 10,000 synaptic connections with other cells. Given their small size, complexity, and inaccessibility, we know far less about how brain synapses form. They are, however, believed to assemble in basically the same way as those at the neuromuscular junction.

In all organisms, synapse formation begins during embryonic development and continues during the early postnatal period. In humans, functional synapses have been observed at 23 weeks of gestation. The few postmortem studies that have been carried out so far suggest that synapses form at different rates within different brain regions, but that typically the number of synapses in most or all regions peaks during the first year of life. In the visual cortex, for example, the formation and stabilization of synapses is highly dependent upon visual experience (see below), and the number of synaptic connections reaches its highest density at between 2.5 and 8 months of age. By contrast, some regions of the developing frontal cortex continue to create new synapses well into the third year of life.[8]

## Synaptic Pruning

Unwanted neuronal connections are eliminated from the developing nervous system by a process called synaptic pruning. Until relatively recently, it was widely believed that synaptic pruning in the cerebral cortex occurs mostly at puberty and is completed during early adolescence. In the past few years, however, it has become clear that synaptic pruning in the prefrontal cortex continues well into the third decade of life before the total number of synapses in the brain stabilizes to adult levels.[9]

Thus, while the human brain reaches its full size by about 16 years of age, the prefrontal cortex does not reach full maturity until this pruning is complete, and these gradual brain changes are associated with changes in behavior. The frontal cortex is associated with complex functions such as decision-making and evaluation of rewards and, because it takes so long to reach full maturity, adolescents tend to place great emphasis on gaining approval from their peers, and often engage in risky behavior to do so. As synaptic pruning refines the prefrontal circuitry during the second and third decades of life, the executive functions improve, and adults behave more responsibly.[10]

Synapse formation and pruning occur extensively in the embryonic brain, and are vital for its proper development. Yet, neither process is restricted to development: the adult brain continues to create new synapses, and to eliminate unwanted ones, throughout life, and we now know that both these processes play important roles in learning, memory, and other aspects of normal brain function (see chapter 4).

## Sensory Experience and Critical Periods

Much of our understanding of how sensory experience shapes developing neural circuits comes from another classic set of experiments, performed by the physiologists

David Hubel and Torsten Wiesel in the 1960s. Using microelectrodes to examine the properties of cells in the primary visual cortex of cats, they had identified neurons that responded highly selectively to visual stimuli consisting of dark bars moving in specific directions.[11] They went on to show that these orientation-selective cells are arranged in alternating columns which respond preferentially to visual inputs from one eye or the other.[12] These ocular dominance columns give the primary visual cortex its characteristic striped appearance, and one of its other names, the striate cortex.

Thus, inputs from the left and right eyes converge in the primary visual cortex, and compete for space there, and in another set of experiments Hubel and Wiesel showed how this competition is driven by visual experience. They reared newborn kittens with one eyelid sutured shut, and found that this had a dramatic effect on development of the visual cortex. As a result, the ocular dominance columns that would normally receive inputs from the closed eye failed to develop, whereas those receiving inputs from the open eye grew to be far larger than they should. Importantly, though, the experiments also showed that the effects were reversible, but only if the eye was reopened before the kittens reached a certain age.[13,14]

This marked another significant advance in our understanding of neural development. It showed that proper development of the visual cortex is highly dependent upon visual stimulation, and established the critical period—a narrow developmental time window during which the nervous system is especially sensitive to particular environmental stimuli—as a key concept not only in developmental neuroscience but also in psychology.

This work—for which Hubel and Wiesel were subsequently awarded the Nobel Prize—also led to an effective treatment for amblyopia (lazy eye), an eye condition that affects about 4% of children. Amblyopia occurs because of improper eye development, and results in reduced vision, misaligned eyes, and poor depth perception. It can be treated by patching up the other eye, which forces the child to use the lazy eye and thus drives the development of that part of the visual pathway. The best outcome is achieved if the treatment is started before 8 years of age.

Subsequent research has shown that the other sensory systems are similarly dependent upon experience for their development. It has also revealed that the timing of the critical period for plasticity in the visual cortex is controlled by the maturation of inhibitory interneurons. Interneurons typically have short fibers that are restricted to single regions of the brain; they synthesize and release the neurotransmitter gamma-aminobutyric acid (GABA), which inhibits neuronal activity. They play important roles in integrating information and regulating the activity of neuronal networks.

The brain contains various types of interneurons, but many of these have not yet been properly characterized, and we probably still do not fully appreciate their diverse forms and functions. But one type in particular—the large basket cells—are evidently responsible for plasticity in the developing visual system.

Large basket cells are present in the primary visual cortex, but they mature slowly. When newborn mice first open their eyes, a protein called Otx2 is transported along the optic nerve from the retina to the visual cortex, where it accumulates inside the large basket cells. At this stage, the large basket cells are still immature, forming numerous weak inhibitory connections with their neighboring neurons. When the concentration of Otx2 reaches a certain level, the molecules enter the nucleus, where they activate a genetic program that promotes maturation of the large basket cells.[15]

As this program unfolds, the large basket cells begin to refine their connections. Certain synapses are stabilized and strengthened while others are eliminated by pruning. Meanwhile, the maturing network of large basket cells is gradually ensheathed by a net of extracellular matrix proteins, which strengthens the new synaptic connections further. Thus, sensory experience refines the microscopic structure of the visual cortex by driving maturation of the large basket cells, which puts the brakes on plasticity by

Sensory experience refines the microscopic structure of the visual cortex by driving maturation of the large basket cells, which puts the brakes on plasticity by consolidating the emerging circuitry at a time when its representation of the world is most accurate.

consolidating the emerging circuitry at a time when its representation of the world is most accurate.[16]

In keeping with this idea, deleting one of the genes needed for GABA synthesis, or administering a drug that blocks or reduces GABA-mediated inhibition, prevents the experience-dependent plasticity of ocular dominance columns in mice. Similarly, an infusion of brain-derived neurotrophic factor (BDNF), a growth factor needed for the survival and maturation of large basket cells, accelerates closure of the critical period. Conversely, when an enzyme that breaks down the extracellular net is injected into the mouse brain, it reopens the critical period; and the transplantation of immature interneurons into the brains of newborn mice induces a second period of plasticity corresponding with maturation of the transplanted cells.[17]

Thus, "critical periods" are not as critical as we once thought they were. The surprise discovery that the timing, control, and closure of critical periods is dependent upon the maturation of long-range inhibitory circuits immediately suggested ways in which they might be "reopened" in later life. Indeed, clinical trials are now under way to test whether drugs that block GABA-mediated inhibition might benefit adults with amblyopia by restoring plasticity in the visual cortex.[18]

# SYNAPTIC PLASTICITY

Nerve cells are specialized to process information, and to communicate with one other, using the language of electrochemistry. They produce electrical impulses that encode information and carry them along their slender fibers, relaying these signals to each other by means of chemical messengers. Synapses are the junctions between nerve cells where this signaling (neurochemical transmission) takes place, and *synaptic plasticity* refers to the various ways in which synapses can be modified.

Most neurons have multiple dendrites, or branches, and a single axon. The dendrites receive signals from other cells, and begin to process them locally, before passing them on to the cell body. Here, the incoming signals are summated; a response signal is then generated at the initial axon segment close to the cell body and propagated along the axon to the nerve terminal. Nervous impulses cannot cross the synapse, and so when an impulse reaches the terminal, it is converted into a chemical signal.[1]

### Functional Architecture of Brain Synapses

Synapses have two structural and functional components, termed the pre- and postsynaptic membranes, which send and receive chemical signals, respectively. Neurons can form synapses with non-neuronal elements, such as skeletal muscle fibers and hormone-producing glands, the so-called "effector" organs. In the brain, however, nerve cells form connections exclusively with one another, with the nerve fiber terminal of one cell coming into close apposition with the axon, dendrite, or cell body of another.

Nerve terminals are often referred to as synaptic boutons, and the postsynaptic elements of excitatory synapses are arranged within tiny protuberances called dendritic spines, whereas those of inhibitory synapses are located in specialized areas of the postsynaptic membrane, found either on the dendrite shaft itself or around the cell body.[2] The synaptic cleft, the minuscule gap between the bouton and spine, is just 20 to 40 nanometers (nm, or billionths of a meter) wide. But despite being so small, synapses are highly organized three-dimensional structures—the boutons and spines are highly specialized to perform their functions, and the behavior of their respective components is tightly orchestrated.

Broadly speaking, there are two types of synapses in the brain: excitatory synapses release the neurotransmitter glutamate, which increases the probability that the

Despite being so small, synapses are highly organized three-dimensional structures—the boutons and spines are highly specialized to perform their functions, and the behavior of their respective components is tightly orchestrated.

postsynaptic cell will generate a nervous impulse, and inhibitory synapses, which use the transmitter gamma-aminobutyric acid (GABA), and this decreases the probability that the postsynaptic cell will fire.

In resting nerve cells, neurotransmitter molecules are stored in tiny, spherical, membrane-bound structures called synaptic vesicles, which are "docked" at the "active zone" just beneath the terminal membrane, awaiting the arrival of a nervous impulse. When an impulse reaches the terminal, it causes an influx of calcium ions through the presynaptic membrane, which in turn causes some of the vesicles to fuse with the membrane and release their contents into the synaptic cleft. Once released, the transmitter molecules diffuse across the cleft, and then bind to receptor proteins embedded in the postsynaptic membrane, triggering it to generate its own impulses. This process is said to be "quantized," as each vesicle contains a specific amount of transmitter molecules, and thus constitutes a quantum (meaning a "packet") of the transmitter.[3]

Neurotransmitters released in this way bind to receptor molecules embedded in the postsynaptic membrane. Some of these receptors are ion channels, which form pores spanning the postsynaptic membrane, and these open upon binding, allowing electrical current—in the form of positively charged sodium, potassium, or calcium ions, or negatively charged chloride ions—to traverse the membrane, altering its conductivity. Others are coupled to

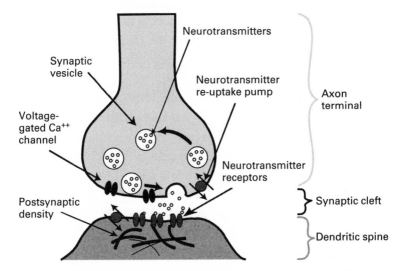

**Figure 3** Pre- and postsynaptic components of a synapse (https://commons
.wikimedia.org/wiki/File:Synapse_Illustration2_tweaked.svg, CC BY-SA 3.0).

so-called second messenger cascades, downstream pathways of enzymes and other proteins, and binding of a transmitter to these receptors brings about longer-lasting biochemical changes within the postsynaptic cell.[4]

In the postsynaptic cell, the movements of neurotransmitter receptors and the various components of their downstream signaling cascades are regulated by an intricate network of scaffolding proteins called the postsynaptic density (PSD), which can be seen with an electron microscope as a fuzzy thickening immediately beneath the membrane. The PSD consists of dozens of different proteins, all of which cooperate to control the movements of receptors and their related molecules within the postsynaptic cell.[5]

Of all the known forms of neuroplasticity, one form of synaptic plasticity, called long-term potentiation (LTP), is the most intensively studied and, therefore, the best understood. LTP is a process that increases the efficiency of synaptic transmission, which is now widely believed to be the neural basis of most, if not all, forms of learning and memory. Modification of synapses also plays an important role in addiction, a maladaptive form of neuroplasticity that involves aberrant learning (see chapter 8).

## Long-Term Potentiation and Long-Term Depression

The idea that memory formation involves the modification of synaptic connections is more than 200 years old. In their correspondences during the 1780s, the Swiss naturalist Charles Bonnet and the Italian anatomist Michele Vincenzo Malacarne discussed the idea that mental exercise can induce brain growth. Malacarne agreed to test the idea by taking pairs of dogs and birds and training one from each pair. A few years later, he dissected the animals' brains, and found that the trained animals had more folds in their cerebella than the untrained ones.[6]

Nearly one hundred years later, the philosopher Alexander Bain suggested that "for every act of memory, every exercise of bodily aptitude, every habit, recollection, train of ideas, there is a specific grouping or coordination of sensation and movements, by virtue of specific growths in the cell junctions."

In the 1940s, the Canadian psychologist Donald Hebb noticed that the lab rats he took home as pets for his children outperformed others on problem-solving tasks when returned to the lab several weeks later. This seemed to show that early experience can have dramatic and permanent effects on brain development and function. Hebb reported these findings in his influential 1949 book, *The Organization of Behavior*, concluding that "the richer experience of the pet group... made them better able to profit by new experience at maturity—one of characteristics of the 'intelligent' human being."

In that book, Hebb postulated that memories are formed by the strengthening of synaptic connections. "Let us assume that the persistence or repetition of a reverberatory activity (or 'trace') tends to induce lasting cellular changes that add to its stability," he wrote. "When an axon of cell A is near enough to excite a cell B and repeatedly or persistently takes part in firing it, some growth process or metabolic change takes place in one or both cells such that A's efficiency, as one of the cells firing B, is increased." In other words, neurons that fire together, wire together.

The idea was way ahead of its time—it was not until nearly 25 years later that Timothy Bliss and Terje Lømo observed a mechanism just like the one Hebb had described. Working on anesthetized rabbits, Bliss and Lømo used microelectrodes to electrically stimulate fibers of the perforant path while simultaneously recording the electrical responses of neurons in the dentate gyrus of the hippocampus, which are at the end of that pathway.

Stimulation of the perforant path fibers evoked an electrical response in dentate gyrus cells, as expected. Bliss and Lømo also found, however, that repetitive stimulation of the fibers (with a frequency of between 10 and 20 Hertz, or pulses per second) caused a massive increase in the size of the electrical response in the dentate gyrus. As well as being far larger, the responses also lasted longer, so that the cells took much longer to return to baseline.[7]

Repetitive stimulation had dramatically increased the efficacy of the neurochemical signaling between the perforant path fibers and neurons in the dentate gyrus, strengthening the synaptic connections between them. In Bliss and Lømo's initial experiments, this strengthening lasted for periods of between 30 minutes and 10 hours, and so they named it long-term potentiation (LTP); but we now know that it can persist for days or weeks, and perhaps even longer.

The induction of LTP is dependent upon binding of the excitatory neurotransmitter glutamate to

*N*-methyl-D-aspartate (NMDA) receptors. The NMDA receptor is an ion channel that is permeable to sodium, potassium, and calcium, but the central pore that allows these ionic currents to pass is blocked by a magnesium ion.

Under normal circumstances, this magnesium block remains, and the glutamate released from a nerve terminal works through two other receptor types, the AMPA and kainate receptors. High-frequency stimulation of the kind that induces LTP increases the amount of glutamate released by the nerve terminal and removes the magnesium block, allowing currents to flow through the NMDA receptors. The influx of calcium is particularly important, as it triggers various enzymes needed for the cellular processes underlying LTP.[8]

NMDA receptors thus have unique biophysical properties that make them perfectly suited to triggering LTP. The magnesium block ensures that they are activated only in response to high-frequency stimulation from the presynaptic cell, and the calcium currents that flow through them are highly localized, producing very discrete "microdomains" of elevated calcium-ion concentration, such that LTP can be restricted to individual dendritic spines, or subsets of them, on a given neuron.[9]

LTP involves changes in both the pre- and postsynaptic components of the connection that is being strengthened. At the nerve terminal, each active zone typically has a pool of several hundred vesicles, but only a small proportion of these are available for release at any time.

High-frequency stimulation enhances glutamate release at the nerve terminal, either by increasing the number of vesicles that fuse with the membrane, expanding the pool of available vesicles, speeding up the recycling process, or a combination of these.

Using methods such as confocal microscopy, it is now possible to tag individual receptor molecules with fluorescent marker molecules or quantum dots, then visualize their distribution and track their movements in living cells, isolated from the brains of animals and kept alive in Petri dishes. Using such methods, researchers have shown that neurons have mobile and immobile pools of glutamate and GABA receptors on their surface, and that receptor molecules can move rapidly around inside neurons.

This receptor trafficking can enhance the responsiveness of the postsynaptic cell. Induction of LTP mobilizes AMPA receptors, inserts them into the membrane, and then shuttles them within it so that they become highly concentrated at the synapse but not in other parts of the dendritic spine. In the same way, LTP is thought to awaken "silent" synapses by insertion of AMPA receptors, which they normally lack. Mobilized receptors are transported in spherical, membrane-bound structures that resemble synaptic vesicles, and are inserted into the membrane by exocyctosis, the same process by which vesicles fuse with the presynaptic membrane during neurotransmitter release.[10]

At excitatory synapses, the movements of AMPA receptors are orchestrated by the scaffolding proteins of the postsynaptic density, which is restricted to the tip of the dendritic spine and which keeps the receptors and their downstream signaling partners anchored in their proper place. Following induction of LTP, calcium currents that flow in through NMDA receptors activate enzymes that redistribute the receptors by rearranging the scaffolding.[11]

Once LTP has been induced, the postsynaptic cell sends a signal back to its presynaptic partner. Once it enters the presynaptic cell, this back-propagated signal activates genes that synthesize the many cellular proteins needed to maintain LTP. The gaseous neurotransmitter nitric oxide has been implicated as this so-called retrograde messenger.

All of these mechanisms are reversible. The rate at which a nerve terminal recycles spent synaptic vesicles can be reduced, for example, leading to a depletion in the number of readily available vesicles at the active zone. And receptors can be removed from the postsynaptic membrane just as rapidly as they can be inserted. Together, these events have the opposite effect of long-term potentiation: they make neurotransmission less efficient, and therefore weaken synaptic connections, in a process referred to as long-term depression (LTD). LTD is also dependent upon the NMDA receptor, but it is induced by repetitive low-frequency stimulation of a presynaptic neuron in the absence of a postsynaptic response.[12]

Bliss and Lømo concluded their classic 1973 paper describing LTP on a cautionary note: "whether or not the intact animal makes use [of LTP] in real life... is another matter." But the fact that it was discovered in the hippocampus, which by then was already strongly implicated in memory, strongly suggested that LTP underlies learning, and ever since then, evidence that synaptic strengthening is indeed necessary and sufficient for memory formation has been slowly accumulating.

For example, when mice are placed into a circular pool of water, they can locate submerged platforms, and quickly form spatial memories of their exact locations, so that they can swim directly to the platforms when placed back into the water later on. But treating the mice with NMDA receptor–blocking drugs during the learning process prevents the formation of spatial memories, so that they are unable to find the hidden platforms afterwards.[13]

Researchers now have more sophisticated methods at their disposal, and one method in particular—optogenetics—allows for the control of neuronal activity with unprecedented precision. Optogenetics involves introducing the genes encoding algal proteins called channelrhodopsins into specific types of neurons. The cells then use their new genes to synthesize channelrhodopsin protein molecules, and insert them into the membrane, making the cells sensitive to light. The cells can then be switched on or off on a millisecond-by-millisecond timescale, depending on which channelrhodopsin they are synthesizing.

Using this method, researchers can now label the hippocampal neurons that fire during memory formation, and reactivate them with pulses of laser light delivered into the animals' brains by optical fibers. Reactivation of hippocampal neurons that fire when mice learn to associate an unpleasant experience with particular location of their environment produces fear responses in the animals, strongly suggesting that the reactivation leads to retrieval of the fearful memories. This same method can be used to manipulate memories in various ways—to switch fearful memories into pleasant ones, or vice versa, and to implant completely false fearful memories into the mouse brain.[14]

Studies such as these provide the most compelling evidence yet that synaptic modification is the neural basis of learning and memory, and it is now widely believed that both strengthening and weakening of synapses are essential for both processes. Current thinking holds that memories form when specific sets of synapses are strengthened and others weakened, within a distributed network of hippocampal neurons, and that retrieval requires reactivation of the same neuronal network.

## Synapse Formation

LTP is a form of functional plasticity that involves transient molecular changes on both sides of the synapse, but learning and memory also involve structural changes that can significantly alter neuronal architecture. As well as modifying the strength of existing synaptic connections, experience and learning lead to the creation of entirely new synapses.

The vast majority of excitatory neurotransmission in the brain takes place at dendritic spines, and so researchers have focused their attention on understanding how learning and experience alter the form of these tiny structures. Dendritic spines were discovered by Cajal over a century ago, in the cerebellum of birds, but it was not until the development of electron microscopy in the 1930s that researchers could study them in any great detail.[15] By cutting brain tissue into a series of ultrathin slices, imaging each one, and then painstakingly reconstructing all the images, they began to get a better idea of how spines and synapses are arranged on the dendrites of postsynaptic neurons, and also of how they can be rearranged in response to sensory experience.

Early studies yielded conflicting evidence. Some showed that spines increase in size by about 15% within 2 to 6 minutes of induction of hippocampal LTP, and then grow even larger at between 10 and 60 minutes, while others showed that LTP causes a marked increase in the surface area of the postsynaptic density. Some researchers observed an increase in the number of spines and synapses, but no changes in size, following LTP induction, and yet others noted significant increases in spine volume but not numbers.[16]

The development in the 1990s of high-resolution time-lapse imaging techniques such as two-photon laser scanning microscopy enabled researchers to examine these processes in even greater detail. Initially, experiments like this were performed in brain tissue dissected from animals and maintained in Petri dishes, but they can also be done in live animals through "cranial windows," or thinned sections of the animal's skull. Combined with the use of sensor molecules, which fluoresce in response to the localized increases in calcium ion concentration produced by NMDA receptor activation, *in vivo* imaging can be used to monitor these processes for prolonged periods of time during sensory experiences or learning of a new motor skill.

These newer methods confirm earlier findings, showing again that sensory experience can produce structural changes to dendritic spine morphology, and that LTP can induce rapid changes in the size, shape, and number of synapses. Following induction of LTP, new spines form on the dendrite, sometimes forming connections with the same synaptic bouton that triggered their formation. The heads of existing spines grow larger, while their necks become shorter and wider. Spine head volume can increase threefold within one minute of repeated electrical stimulation. All of these changes facilitate the trafficking of receptors into the spine heads, making them more sensitive to glutamate.

Learning and experience likely lead to the patterned formation of new spines along the same dendrite branch and also across other branches of the same dendritic tree. Motor learning induces clusters of new spines to form in adjacent locations on the dendrites of cells in the mouse motor cortex, and causes weakening and shrinking of neighboring clusters; the new clustered spines are more persistent than spines that form alone.[17]

It is tempting to speculate that the persistence of memory is related to the stabilization of new dendritic spines and to synchronized activity in neighboring synapses. Structural changes to dendrites involve reorganization of the filamentous proteins that make up the postsynaptic density, by the same signaling pathways triggered by the NMDA receptor following induction of LTP. Furthermore, different motor tasks activate NMDA receptors to produce calcium microdomains on different branches of individual pyramidal neurons in the mouse motor cortex. Thus, individual branches of dendritic trees, or subsets of spines on them, might serve as basic units for storing information. Such mechanisms could help to explain the brain's extraordinary capacity for memory storage.[18]

Yet, the precise relationship between synaptic modification, spine formation, and memory is still unclear, and there is some evidence to suggest that new spines are not actually necessary for memory. For example, spine density in the squirrel brain decreases dramatically during hibernation and increases again afterwards, but the animals can still remember tasks they learned before they started hibernating. Similarly, spine density in the hippocampus is reduced by 30% in female rats in estrus, but they can still remember items they learned earlier in the menstrual cycle.

Findings like this suggest that the persistence of dendritic spines is not necessary for long-term storage of memories. But the conflicting findings on exactly how experience and learning alter dendrite architecture could be due in part to differences in the type of stimulation used, or the brain area being studied. There is even some evidence suggesting that merely handling brain tissue in preparation for experiments can alter the density of spines within it.

To complicate matters, dendritic spines exist in a variety of forms, and its thought that any individual spine can morph between and adopt all of them. There are mushroom-like spines with large, round heads attached to their parent dendrite by a narrow neck; long spines, which appear as thin, finger-like protuberances; and small spines, which are short and stout and have no noticeable neck. It's possible that each of these forms contributes to different aspects of memory storage, or that different types of memory cause different types of structural changes to dendrite architecture.[19]

Synapses can also be weakened, and the spines associated with them can shrink, pull away from their presynaptic partners, or even retract and be eliminated altogether. Synapse elimination, or synaptic pruning, occurs extensively during brain development, and is critical for shaping and fine-tuning neural circuits as they form (see chapter 3). Pruning also takes place widely in the adult brain, and, like LTP and synapse formation, is thought to be necessary for learning and memory.

Thus, learning, memory, and other experiences probably produce widespread patterns of synaptic modification throughout entire networks of neurons in particular regions of the brain, depending on the type of experience. Synaptic modification takes place continuously throughout the brain, and it is likely that millions of synapses are modified in the human brain every second in one way or another. Current imaging methods are rather limited in their field of view, being restricted to several branches of a dendritic tree, but emerging techniques such as superresolution microscopy will undoubtedly reveal more about dendritic spine dynamics and their contribution to long-term memory.

## Glial Cells: Partners in Plasticity

Glial cells are the nonneuronal cells of the nervous system, outnumbering neurons by about ten to one. They were discovered at around the same time as neurons, but were believed to play only supportive roles such as providing nutrition and insulating nerve fibers; hence their name *glia*, meaning "glue." Glial cells do perform these roles, but we now know that they also make important contributions to—and are just as crucial for—information processing in the brain and spinal cord.

Traditionally, synapses were thought to consist of just two elements, the presynaptic bouton and postsynaptic membrane. In the early 1990s, however, evidence began to emerge that they are in fact tripartite structures, and that glial cells called astrocytes regulate the chemical signals that are transmitted between neurons.

Astrocytes are star-shaped cells that were initially thought to fill the extracellular spaces in brain tissue. But it is now clear that they not only respond to neuronal activity but can also produce their own electrical signals, and they synthesize and release a whole host of neurochemical transmitters, including glutamate and GABA.

Astrocytes are by far the most numerous cell type in the brain. Each one has many fine branches that come into contact with hundreds of dendrites and up to 150,000 individual synapses. These processes are highly motile, and rapidly extend toward and envelop active synapses. Electron microscopic examination of brain tissue reveals that their fibers interact with large dendritic spines in response to neuronal activity, and that these fibers are less motile than those associated with small spines.

Large spines tend to be more persistent than smaller ones, and so it seems that astrocytes help to stabilize those spines with active synapses. There is also some evidence that astrocytes can modulate synaptic signaling by clasping synapses to restrict the diffusion of neurotransmitters, or loosening their grip to allow them to flow more freely.

Astrocytes form networks with each other and with their neuron neighbors. Whereas neurotransmission takes place over a timescale of milliseconds, astrocyte activity lasts for a few seconds. When an astrocyte releases glutamate, it excites whole clusters of neurons, and their prolonged activity may be a way of synchronizing activity of entire populations of neurons. The prolonged activity of astrocytes may also contribute to LTP by persistently activating postsynaptic membranes to coincide with incoming signals.[20]

Microglial cells also play important roles in synaptic plasticity. These are the brain's resident immune cells, which provide the first line of defense against infection and injury. They are deployed to damaged sites, where they engulf pathogens and cellular debris by enveloping them in a small segment of membrane and then internalizing them, a process called phagocytosis, or "cell eating."

It turns out that the developing brain treats unwanted synaptic connections in exactly the same way. Unwanted connections are "tagged" for destruction with immune system molecules called complement proteins. Microglia

Unwanted connections are "tagged" for destruction with immune system molecules called complement proteins. Microglia recognize this as a signal saying "eat me," and engulf all the tagged synapses they come across.

recognize this as a signal saying "eat me," and engulf all the tagged synapses they come across. It's now thought that microglia are responsible for synaptic pruning throughout the developing brain, as well as for the extensive pruning that occurs in adolescence (see chapters 3 and 9).

Synapses are also constantly being eliminated in the adult brain, and it seems that microglia are responsible for this, too. They continuously patrol their patch of brain tissue, and preferentially contact stubby spines, which are usually the least persistent of newly formed spines. Thus, microglia seem to monitor the status of synapses in their patch and engulf the unwanted ones.[21]

# ADULT NEUROGENESIS

The fine structure of the nervous system was a subject of heated debate throughout the nineteenth century. In the late 1830s, after viewing plant and animal tissues under the microscope, the German scientists Theodor Schwann and Matthias Schleiden proposed that cells are the basic building blocks of all living things, a view that came to be called the cell theory. But the microscopes available at the time were not powerful enough to resolve synapses, which measure approximately 20 to 40 nanometers (nm, or billionths of a meter), and so it remained unclear whether this also applied to the nervous system.

Investigators were split into two camps. Some believed the brain and spinal cord consisted of a reticulum, or a continuous network of tissue, while others argued that the nervous system, like all other living things, must be made up of cells. With improvements in microscopy and the methods for staining and visualizing their samples, investigators could view nervous tissue in increasing detail, and by the turn of the twentieth century, the long-standing debate was finally settled.[1]

Thanks largely to the work of Cajal, researchers came to accept the so-called neuron doctrine, which stated that specialized cells called neurons are the basic structural and functional units of the brain and spinal cord. Cajal and others had studied how the nervous system of man and other animals develops, and described the various stages through which neurons pass as they are maturing: birth by cell division, followed by migration of the daughter cells, growth and extension of their fibers, and, finally, the precise formation of synaptic connections. Because they never saw immature neurons in adults, they concluded that brain structure becomes fixed soon after birth.

In his 1913 book, *Degeneration and Regeneration of the Nervous System*, Cajal stated that the neural pathways in the adult brain and spinal cord are "something fixed, ended, and immutable." This conclusion came to be widely accepted, and before long, the idea that the adult mammalian brain does not create new cells became a central dogma of modern neuroscience. Most researchers agreed that while vast amounts of neurons and glial cells are generated during development, this process ends in the period just after birth. Thus, it followed that we are born with all the brain cells we will ever have, and that those that are lost through injury or disease can never be replaced.

This dogma persisted for the best part of a century, even though evidence challenging the idea began to emerge in the early 1960s, following the introduction of a technique called [³H]-thymidine (or tritiated thymidine autoradiography. In this process animals are injected with radioactive thymidine, which is taken up by cells and incorporated into the newly synthesized DNA found in newborn cells. Their brains are then dissected, and X-rays are used to detect any radioactivity.[2]

Joseph Altman and Gopal Das of the Massachusetts Institute of Technology began using this technique to examine various animal species, and they soon published evidence of the growth of new brain cells in the dentate gyrus, olfactory bulb, and cerebral cortex of the rat, and also in the cortex of the cat.[3] These initial findings were independently replicated and confirmed by others in the early 1980s, but they were met with skepticism by the scientific community, and largely ignored.[4,5]

Soon, more evidence came from the brains of songbirds. Sexually mature male canaries learn a new song every year, in order to serenade potential mates, and learning and production of their songs are controlled by two brain nuclei. Fernando Nottebohm of Rockefeller University performed a series of experiments which showed that the size of these brain nuclei fluctuate with the seasons: both were found to be substantially larger in the spring than in the fall.

Nottebohm hypothesized that these fluctuations were due to an increase and then a reduction in the number of synapses and neurons within the song-producing nuclei. When mating season ends, large numbers of neurons die off, causing the nuclei to shrink; in the spring, however, the nuclei are regenerated by the production of new neurons, so that the bird can learn to sing once again. Nottebohm had not only discovered a clear and direct link between brain and behavior, but his results also "showed beyond reasonable doubt that neurons are born in adulthood and incorporated into existing circuits."[6,7]

A series of advances and discoveries finally broke the long-standing conviction that the mammalian brain lacks the ability to regenerate itself. In the late 1980s, Elizabeth Gould and her colleagues at Princeton University began to publish evidence of newborn neurons in the hippocampus of adult rats and then, some time later, in both the hippocampus and cerebral cortex of macaque monkeys. Evolutionarily, monkeys are far more closely related to humans than rats are, and so this raised hopes that the human brain may also continue to form new cells throughout life.[8]

The development of new techniques using fluorescently labeled antibodies that bind to specific cellular proteins enabled researchers to distinguish between the neurons and glial cells in their tissue samples. In 1992, a pair of researchers at the University of Calgary in Alberta, Canada used these methods to identify and then isolate neural stem cells from the brains of adult mice.[9] Neural stem cells are said to be "multipotent" because they retain their undifferentiated, embryonic state and can go on to form any type of cell found in the brain. But they divide asymmetrically, so, while they give rise to new neurons and glial cells, they can also renew themselves indefinitely.

Subsequent research revealed that the brains of adult mice and rats contain two discrete populations of neural stem cells. During early development, the nervous system consists of a hollow tube running along the back of the embryo, and the inner lining of this neural tube is packed with stem cells, which divide to produce immature neurons that migrate through the thickness of the tube. At the front end of the tube, successive waves of migrating cells jostle past each other to form the layers of the cerebral cortex, one after the other, from the inside out. Further back, smaller numbers of cells migrate outward to form the spinal cord.

In adults, neural stem cells are restricted to two discrete niches within the walls of the lateral ventricles: the subventricular zone, which creates cells that migrate through the rostral migratory stream to the tip of the olfactory bulb, and the dentate gyrus of the hippocampus, whose new cells stay near their birthplace and differentiate into granule neurons.[10]

The neurons formed in these niches appear to be critical for brain function and behavior. Experiments using genetic engineering to kill off newly generated cells as soon as they are born, or at a specific time point in the animals' lives, show that the addition of new neurons to the olfactory bulb is essential for the formation of new smell memories, while those added to the hippocampus contribute to spatial memory, object recognition, and pattern separation, the process by which the brain distinguishes between similar patterns of neural activity.[11]

Certain environmental factors can regulate the process to dramatically affect the rate at which new neurons are produced. For example, physical activity, environmental enrichment, and learning tasks enhance the proliferation of neural stem cells and, in some cases, promote the survival of newborn neurons, whereas stress, certain types of inflammation, and sensory deprivation have the opposite effect.[12]

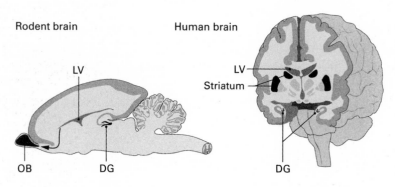

**Figure 4**  Neurogenic niches in the rodent and human brains.

Physical activity, environmental enrichment, and learning tasks enhance the proliferation of neural stem cells and, in some cases, promote the survival of newborn neurons, whereas stress, certain types of inflammation, and sensory deprivation have the opposite effect.

Another major breakthrough came in 1998 with the publication of a landmark study that provided the very first evidence that the human brain also forms new cells throughout life. The late Peter Eriksson and his colleagues realized that doctors were injecting cancer patients with bromodeoxyuridine (BrdU) in order to visualize and monitor the growth of their tumors. BrdU, like tritiated thymidine, is an analog of one of the four chemical bases found in DNA; as such it is incorporated into the newly synthesized DNA of newborn neurons. Eriksson and his colleagues got permission to examine the brains of five such patients after they had died. They treated samples of hippocampal tissue with different fluorescently labeled antibodies that bind to BrdU and to proteins expressed by neurons but not by glial cells, and detected newborn neurons in all five samples.[13]

Eventually, neural stem cells were isolated from the human brain, too. In rodents, these two populations of neural stem cells continue to generate new cells throughout life but the rate at which new neurons are produced decreases with age; the same pattern is also seen in the human hippocampus.[14]

There are important differences, however. The rostral migratory stream is found in the human brain, and even has a unique "ribbon" that branches off toward the frontal cortex. But evidently this pathway is active only up to early childhood. Extensive migration takes place until about 18 months of age, but subsides in older children and is almost completely absent in adults. In this respect, humans appear to be unique among the mammals.[15,16]

By contrast, a 2013 study by researchers in Sweden shows that the human hippocampus produces about 700 cells per day—which corresponds to an annual turnover of about 1.75% of the total number of cells in that part of the brain—and that the rate decreases only slightly with age.[17] More recently, the same group published evidence of adult neurogenesis in the human striatum, a subcortical structure involved in motor control, reward, and motivation. These cells apparently originate in the subventricular zone and go on to form interneurons, whose fibers are restricted to the immediate area and whose inhibitory signals are vital for circuit function.[18]

The vital question is, does adult neurogenesis in the human brain serve any purpose, as it does in birds and rodents? The extent of adult neurogenesis in the human hippocampus is comparable to that seen in rodents, so it's certainly possible that the new neurons contribute to brain function, but there is as yet no direct evidence for this.

In adult mice, fluoxetine (Prozac) and related antidepressants stimulate hippocampal neurogenesis. This finding led some researchers to speculate that neurogenesis may play a critical role in the development and treatment of depression. Animal experiments in which adult hippocampal neurogenesis is eliminated altogether have produced conflicting results: in some, the animals display an enhanced stress response and increased depression-like behaviors, but in others they do not.

In humans, depression is indeed associated with a reduction in hippocampal volume, but it is by no means clear that this reduction is due to impaired neurogenesis. It's possible that impaired neurogenesis is one of many factors contributing to this complex disease, and is more important in some cases than in others. Likewise, the study showing that neurons are continuously added to the striatum also showed that adult-born cells are depleted in Parkinson's disease, but it is still unclear whether or how this process is related to impaired neurogenesis.[19,20] There is another possible downside to adult neurogenesis. Cancer arises when cells divide uncontrollably and spread through the body, and so it's possible that the neural stem cell populations present in the adult human brain could contribute to the formation of brain tumors.[21]

Skeptics maintain that the numbers of cells produced by the adult human brain are too small to be of functional significance. They also argue that adding new cells could disrupt the stability of existing neuronal circuitry and, therefore, that the process is nothing more than a relic left over from our evolutionary ancestors.

The most outspoken skeptic is the developmental neurobiologist Pasko Rakic, of Yale University. In the early 1970s Rakic performed a series of highly influential

Skeptics maintain that the numbers of cells produced by the adult human brain are too small to be of functional significance.

studies revealing how immature neurons migrate through the developing monkey brain, and he has worked on monkeys ever since. He has repeatedly failed to find any evidence of adult neurogenesis in the monkey cerebral cortex, and he is critical of the labeling methods used to identify newborn cells.

From his work in monkeys, Rakic estimates that neurons added to the adult human brain would probably take about a year to reach full maturity. This, he says, makes it highly unlikely that Prozac and related drugs work by stimulating neurogenesis, because they take just six weeks to exert their effects. There is some evidence that newborn neurons in the adult brain have enhanced synaptic plasticity, however, and so some argue that a year-long period of immaturity could actually make newborn neurons better able to contribute to brain function.[22]

Despite the controversy, the discovery of neurogenesis and neural stem cells in the adult human brain quickly raised hopes of stem cell–based therapies for neurological injury and disease, and also suggested two potential approaches for how such therapies might be developed. We know that neural stem cells can divide in response to brain injury, suggesting that this self-repair mechanism might one day be harnessed, by coaxing the brain's endogenous stem cells to deploy new cells that would travel to an injury site and replace those that have been damaged or killed. An alternative strategy is to transplant stem cells into the brain and target them to the injury site.

Our understanding of neural stem cell biology is still far from complete, and researchers trying to develop such therapies face major technical challenges. Which types of stem cells are most appropriate for transplantation, and might different types be better suited to a given disease or type of injury? What is the optimum number of cells to be transplanted? And how can we be sure that transplanted cells will survive for long enough to integrate and aid in the recovery of neurological function?

Because of these difficulties, stem cell–based therapies for neurological disease and injury are still far from achieving their full potential, and in fact all of the clinical trials conducted so far have failed.[23] Regardless, public awareness of these issues has led to a dramatic increase in stem-cell tourism to countries whose lax regulations enable unscrupulous vendors to sell unapproved—and possibly dangerous—therapies to desperate patients.

# BRAIN TRAINING

The term "brain training" usually refers to computer games designed to improve mental functions such as attention and working memory. These games are purported to improve such functions—and overall brain health—by exercising the organ, in the same way that physical activity helps to maintain the health of the body.

Today, brain training is a multimillion dollar industry, and there are dozens of companies selling computer games, gadgets, and smartphone applications. Many of these products are targeted toward the aging and elderly, and are purported to improve overall brain health and even reduce the risk of developing Alzheimer's disease and other forms of dementia. Currently, however, there is very little evidence that brain training products lead to what psychologists call transfer effects: they do lead to significant improvements in the mental abilities needed to perform well at the game, as would be expected, but it

Currently, there is very little evidence that brain training products lead to what psychologists call transfer effects.

is still unclear whether they also improve other, unrelated cognitive functions.[1]

In October 2014, a large group of eminent researchers issued a joint statement on the subject. "We object to the claim that brain games offer consumers a scientifically grounded avenue to reduce or reverse cognitive decline when there is no compelling scientific evidence to date that they do," they wrote. "The promise of a magic bullet detracts from the best evidence to date, which is that cognitive health in old age reflects the long-term effects of healthy, engaged lifestyles. In [our] judgement, exaggerated and misleading claims exploit the anxiety of older adults about impending cognitive decline. We encourage continued careful research and validation in this field."[2] About one year later, the San Francisco-based brain training company Lumosity was ordered by the U.S. Federal Trade Commission to pay a $2 million settlement for deceiving customers with unfounded claims about the benefits of their products.[3]

As we have seen, though, the brain is continuously being shaped by our experiences, and there is now plenty of evidence that other types of brain training have significant effects on the organ. Animal research has provided important information about the cellular mechanisms of plasticity induced by training. For example, training rats to make fine-grained time judgments expands the bandwidth sensitivity of auditory neurons, whereas training owl monkeys on a touch discrimination task shrinks the receptive fields of neurons in the primary somatosensory cortex, the part of the brain that processes touch information.

Lasting changes are far harder to study in animals and, until relatively recently, in humans, too. In the past 20 years or so, however, the use of neuroimaging technologies has become widespread, and the number of studies using these methods to investigate the neural consequences of long-term training is growing. With these methods at their disposal, researchers have shown that learning a second language is associated with various anatomical changes in the brain. Similarly, some people spend years or decades acquiring other types of knowledge, skills, or expertise. Such rigorous, long-term training also leads to long-lasting changes in both the structure and function of the brain. Professional athletes, musicians, and the like, are therefore a fascinating natural laboratory for the study of experience-dependent neuroplasticity.[4]

**Language Learning**

A pioneering 2004 study used voxel-based morphometry to examine the brains of bilingual Europeans and compare them to those of monolinguals. This revealed that bilingualism is associated with increased gray matter density in the left inferior parietal lobule, a region of the brain that has been implicated in a number of important language-related functions, such as phonological working memory (or memory for language sounds), lexical learning, and the integration of information from diverse sources, and so the volume increase may reflect acquisition of second-language vocabulary.

The researchers found that the effect was bigger in early than in late learners: participants who started learning a second European language before the age of 5 exhibited larger volume increases than those who learned later. The extent of the change was also related to individuals' ability for language learning, such that those who were more proficient at acquiring their second language showed greater increases in gray matter volume than those who found it more difficult.[5]

Subsequent studies have confirmed these initial findings and have also demonstrated that learning a second language is associated with other kinds of anatomical changes, including changes in cortical thickness in brain areas linked to language, as well as changes in the architecture of the white matter tracts that interconnect them. Even short-term language training alters brain structure: various studies show that college students and military interpreters who enrolled in intensive three-month language courses exhibited brain differences compared to controls who had not.

The anatomical changes associated with language learning appear to be reversible, though. One brain scanning study found that adult Japanese speakers who took a six-week English-language course had increased gray matter density, compared to controls, in certain language regions of the brain. Follow-up scans performed a year later revealed even bigger increases in those who had kept up their language practice. In those who had stopped, however, gray matter density in the affected brain regions had returned to pre-training levels.[6]

Unlike commercially available brain training products, language learning does appear to have transfer effects, and evidence that lifelong bilingualism confers certain advantages is beginning to emerge. Bilingualism requires switching between languages and selecting the correct vocabulary, among other tasks that exercise so-called executive functions such as reasoning, task switching, and problem solving. Furthermore, learning a second language apparently has neuroprotective effects; thus it may reduce the risk of Alzheimer's disease and other neurodegenerative conditions, even when it takes place in later life, by increasing "cognitive reserve"— a somewhat vague term that refers to the mind's resistance to brain damage.[7]

## Musical and Athletic Training

Early neuroimaging studies revealed that long-term training regimes are associated with differences in both gray and white matter. Thus, classical musicians who started training before the age of 7 have a larger corpus callosum than those who started their training later and nonmusical controls. This huge bundle of nerve fibers contains fibers that cross from one side of the brain to the other and coordinates the activity of the limbs.[8] And the dexterity required of professional violinists is associated with extensive reorganization of finger representation in the primary somatosensory cortex. The representation of the fingers of the left hand in the somatosensory cortex was larger in the musicians than in healthy nonmusical controls, and the difference was more pronounced in those who had started their training at an earlier age. Cortical representation of the right hand, with which string instrument players hold the bow, remained unchanged.[9]

More recent studies provide similar findings. Using a neuroimaging technique called voxel-based morphometry, researchers have shown that professional keyboard players have larger gray matter volumes in motor, auditory, and visuospatial brain regions compared to amateur musicians and nonmusical controls, with the extent of the change again correlated to the length of time as a musician.[10]

Others have used diffusion tensor imaging (DTI) to show that piano practice alters the brain's white matter tracts, and that the effects depend on the age period during which the training took place. The changes are seen in the corpus callosum and in fibers that descend from the sensory and motor cortices, and were most pronounced in professional concert pianists who began training before the age of 7.[11] Similarly, karate black belts exhibit significantly larger white matter tracts in the motor cortex and cerebellum than novices and controls, giving them superior motor coordination and enabling them to pack more power into their punches and kicks.[12]

Most of these studies involved recruiting small groups of experts or professionals and comparing the structure or function of their brains with those of amateurs or novices at a single point in time. This cross-sectional experimental design cannot conclusively establish whether any observed differences are the result of training, or whether they reflect anatomical and genetic differences that were present at birth—it may be the case that certain people are born with brains that predispose them to acquiring a particular skill or expertise more easily than others. Distinguishing between these possibilities would require longitudinal studies in which members of each group have their brains scanned repeatedly over a period of months or years.

The few longitudinal MRI studies that have been performed to date do suggest that at least some of the observed differences are indeed due to training. For example, learning to juggle over a period of several months leads to increased gray matter density in the occipitotemporal cortex, which contains motion-sensitive neurons.[13] It also enlarges the white matter tracts beneath the intraparietal

sulcus, which contains brain regions that integrate percep-
tual and motor information and are critical for controlling
and coordinating arm and eye movements.[14]

## The Knowledge

A series of studies performed on London taxi drivers over
the past 15 years provides yet more evidence that mental training can indeed induce anatomical changes to the
brain. To qualify as a licensed London taxi driver, trainees
undergo years of comprehensive memory training to learn
the labyrinthine layout of some 26,000 streets within a
six-mile radius of Charing Cross station, the location of
thousands of landmarks, and also the quickest way to navigate between any two points in the city.

Prospective taxi drivers typically spend three to four
years studying maps and driving around the city, in order
to acquire "the Knowledge" of London's streets. During
this time, they also take a set of stringent examinations
designed to test their spatial learning of each city district,
and are allowed a limited number of attempts at each before progressing on to the next. Only after successfully
completing all of these examinations can they qualify and
obtain a license to operate one of London's famous black
taxis, and approximately half of those who begin the training fail their examinations or drop out at some point.

In 2000, researchers at University College London published a study showing that gray matter density in the posterior hippocampus is significantly larger in qualified London taxi drivers than in controls. This brain structure is known to be involved in spatial navigation, and the study also showed that its size was closely correlated with the amount of time spent as a taxi driver—the more experienced the driver, the larger was their posterior hippocampus.[15]

This study was also a cross-sectional one, so the researchers could not rule out the possibility that the differences they had observed were due to preexisting anatomical differences, but they went on to perform several follow-up studies that confirmed that the changes were indeed due to the prolonged and rigorous training regime. First, they scanned the brains of some London bus drivers, who also navigate London's streets, along far simpler, predetermined routes, revealing that gray matter density in their hippocampi is not significantly different from that of controls.

Next, the researchers carried out a longitudinal study in which they repeatedly scanned the brains of trainee taxi drivers as they underwent training. Of the 79 trainees enrolled in the study, 39 went on to qualify as taxi drivers, and 20 failed the training but agreed to return for brain scanning nevertheless. Those who qualified exhibited the same increases in gray matter density, but the hippocampi of those who had failed looked no different from those of subjects in the control group.[16]

Together, these studies show that the comprehensive memory training required to successfully complete "the Knowledge" induces specific changes in brain anatomy. Just as weightlifting leads to an enlargement of muscle tissue, so too can mental training expand corresponding parts of the brain. This comes at a price, however—qualified London taxi drivers appear to be worse at acquiring new visuospatial information than others, and some researchers suspect that their increasing use of satellite navigation devices could lead to a gradual deterioration of the hippocampus.

Thus, the brain is a highly dynamic organ that adapts to its user's demands. Intensive training alters the brain in such a way that it begins to execute the appropriate functions more efficiently. Musical and athletic training enhance the execution of the complex sequences of movements needed, and trainees acquiring "the Knowledge" learn how to organize huge amounts of spatial information and then use it effectively. In this way, training optimizes the brain areas and neural pathways involved in performing a given task; as a result, the individual's performance on that task improves, and the task eventually becomes automatized and effortless.

The available data suggest that gaining expertise in any domain requires at least four hours of training per day

Training optimizes the brain areas and neural pathways involved in performing a given task; as a result, the individual's performance on that task improves, and the task eventually becomes automatized and effortless.

for approximately 10 years. Remarkably, there is also compelling evidence that motor imagery—that is, visualizing certain movements in the mind's eye—can also enhance the learning and execution of certain skills. Thus, imagined movements appear to be equivalent to those that are actually performed, and merely "going through the motions" in one's mind can lead to the same kind of plastic changes in the brain.[17]

## Of Mice and Men

Neuroimaging studies have provided a wealth of information about how prolonged intensive mental training alters the brain, but they tell us nothing about the molecular and cellular mechanisms underlying the observed changes. Experiments performed on rodents show that rigorous training on motor tasks can exert various cellular effects, such as the sprouting and pruning of new dendritic spines and axonal branches. It is impossible to observe equivalent processes in humans, however, both because the resolution of current neuroimaging technologies is far below that needed to visualize them and because the techniques used in mice and rats cannot be applied to studying the human brain.

Increases in gray matter density and volume could be explained by adult neurogenesis. That explanation is particularly attractive in the case of London taxi drivers, since the hippocampus is currently the only region of the human brain that is known to continue generating new neurons throughout life (see chapter 5). But the increases can also be explained by the formation of new dendritic spines and synapses and the sprouting of new axon branches. Increases in the number of glial cells, or the formation of new blood vessels to supply new structures with blood, could also increase gray matter density.

Likewise, changes in white matter structure could be due to various mechanisms, such as the addition and removal of myelin from axons, or alterations in myelin thickness or in the spacing between nodes of Ranvier, all of which would alter the conducting properties of a neuron. Although diffusion tensor imaging is sensitive to variations in myelin, it is not yet sensitive enough to distinguish between these mechanisms.[18]

Neuroimaging data can sometimes seem counterintuitive and are often difficult to interpret. One recent study compared brain activity of professional soccer players and swimmers while they performed identical foot movements, and found that the soccer players exhibited less activity in the motor cortical area corresponding to the foot than did the swimmers. The researchers interpreted this as meaning that years of training enable the soccer players to control their foot movements efficiently while also conserving their neural resources.[19]

Clearly, the brain is highly flexible, but we are only just beginning to understand the many ways in which it can adapt to the demands placed upon it. Technological advances will allow for increasingly sophisticated ways of imaging the brain, and will surely deepen our knowledge of how different types of training affect brain structure and function.

# NERVE INJURY AND BRAIN DAMAGE

Neuroplastic changes of various kinds occur in response to nerve injury and brain damage caused by stroke and other insults. Nerve injury leads to changes in the damaged nerve fibers, as well as to functional reorganization of neuronal circuitry in both the brain and spinal cord. These effects can persist for many months or years. The changes that occur following nerve injury and amputation do not lead to any significant recovery of function, especially after a serious injury, but rather can be maladaptive, causing neuropathic pain, or the "phantom" sensations and pain experienced by amputees. By contrast, the spontaneous plastic changes that occur after a stroke can help the brain to compensate for the damage that has occurred.

Investigations of injury-induced brain changes are done in rats, monkeys, and humans. In rats, they focus largely on an area of the brain called the barrel cortex, which receives sensory information from the whiskers. In monkeys and humans, studies are focused on the primary somatosensory cortex, which receives sensory information from the skin surface, and on the primary motor cortex, which executes movements by sending commands down the spinal cord to the muscles. Sensory brain regions are said to be organized topographically. For example, touch information from adjacent areas of the skin surface is processed in adjacent patches of the primary somatosensory cortex, while adjacent groups of body muscles are controlled by cells in adjacent patches of the primary motor cortex. In this way, the body is "mapped" onto the surface of the primary somatosensory and motor cortices.

The size of these cortical representations is determined by the number of nerve endings or muscles in the corresponding body part, rather than by its actual size. Therefore, the vast majority of neural tissue in the primary somatosensory and motor cortices is devoted respectively to processing information from, and sending movement commands to, the face and hands, which are the most sensitive and articulated parts of the body. These cortical representations can be altered by experience, shrinking when they are deprived of sensory information or expanding with increased use of their corresponding part. This process, called remapping, takes place after nerve injury and brain damage, and in some cases it can be induced artificially, using various methods of noninvasive brain stimulation, to facilitate rehabilitation.

## Peripheral Nerve Injury

Some of the earliest direct evidence for neuroplasticity came from animal studies of nerve injury, performed in the early 1980s. When the median nerve in monkeys' arms is severed, corresponding regions of the primary somatosensory cortex are deprived of inputs, but those regions do not lie dormant. In the weeks following the injury the primary somatosensory cortex reorganizes itself so that neighboring patches of brain tissue, which receive inputs from adjacent body parts, expand and encroach upon the deprived area.

In rats, severing the sciatic nerve causes a threefold expansion of the somatosensory area normally devoted to inputs from the adjacent saphenous nerve, which begins one to two days after the nerve has been severed and persists for up to six months. And two to eight months after monkeys have a finger amputated, the patch of primary somatosensory cortex that previously responded to the amputated digit responds instead when adjacent digits are touched.[1]

Reorganization of the motor cortex happens along similar lines, but with different consequences. In rats, the facial nerve normally controls whisker movements, and when it is severed, the corresponding motor area initially sits silent and does not respond to electrical stimulation. A few hours later, however, stimulation produces muscle contractions in the forearm and eyelid.

Two to eight months after monkeys have a finger amputated, the patch of primary somatosensory cortex that previously responded to the amputated digit responds instead when adjacent digits are touched.

Researchers can now detect changes such as these taking place in the human brain, using noninvasive brain stimulation techniques such as transcranial magnetic stimulation (TMS) and transcranial direct current stimulation (tDCS). These changes begin to take place minutes after transient nerve blockade, and weeks after spinal cord injury. Immediately after nerve block by local anesthetic, for example, the patch of motor cortex corresponding to the paralyzed area falls silent, and neighboring areas increase their output to the spinal motor neurons. This effect is reversible, however, and disappears about 20 minutes after the anesthetic has worn off.

The same sort of cortical reorganization also takes place following arm amputation. As is the case in animal studies, the somatosensory cortical region corresponding to the amputated arm gradually shrinks, and the surrounding areas begin to expand and encroach upon it. The vast majority of amputees experience phantom limb, the vivid sensation that the missing limb is still attached to their body, which can often be extremely painful. Phantom limbs are believed to occur at least partly because of the cortical reorganization that occurs following amputation. In the somatosensory and motor cortices, the neural representation of the hands lies immediately next to that of the face; after amputation of an arm, therefore, it is the area representing the face that expands and invades its deprived neighbor. Consequently, touching certain parts

Phantom limbs are believed to occur at least partly because of the cortical reorganization that occurs following amputation.

of the amputee's face may trigger vivid phantom sensations, suggesting that the deprived area somehow retains a memory of its previous function.[2]

It is difficult to explain cortical reorganization in terms of cellular mechanisms, however, since brain scanning technologies are still nowhere near sensitive enough to detect such processes in humans. But the animal studies give us a good idea of what might be happening. They show that neurons whose fibers are severed quickly retract their dendrites, so that they become detached from the nerve terminals associated with them and there is an overall reduction in the number of synapses the cell receives. At the injury site, any intact axon fibers may sprout new branches that grow into the damaged area, and the mismatch between the fibers and their new targets can contribute to neuropathic pain.

The earlier stages of cortical reorganization are thought to involve "unmasking" of previously silent connections, including horizontal ones between adjacent patches in corresponding parts of the cortex and vertical ones from the thalamus, an area that relays all types of sensory information from the sense organs to the appropriate area of the cerebral cortex. Unmasked connections are believed to be strengthened by LTP (see chapter 3), but longer-lasting changes are probably consolidated by the sprouting of new axon branches, by elongation and branching of dendrites, and by formation of new synaptic connections. The animal studies suggest that axons and dendrites can grow for distances of up to 3 millimeters during reorganization of the somatosensory cortex, while the boundaries of representations in the motor cortex can shift rapidly by up to 2 millimeters.[3]

## Stroke

Numerous studies have examined the cortical reorganization that takes place after stroke. Stroke is a leading cause of death and disability, and involves an interruption of the blood supply to the brain, due to blocked or broken blood vessels, that leads to death of oxygen-deprived cells. The frontal and temporal lobes of the brain are particularly vulnerable to oxygen deprivation, and so strokes often damage these regions, leading to the characteristic symptoms of speech deficits and muscle weakness or total paralysis one side of the body. Since the left hemisphere of the brain controls the right side of the body, and vice versa, stroke paralyzes the limbs on the side opposite to the damage.

Unlike nerve injury–induced plasticity, which is rarely helpful, the cortical reorganization that occurs after a stroke is believed to contribute significantly to the recovery of motor function. Paralysis occurs because the stroke damage disrupts the neural pathway that descends from the primary motor cortex to the motor neurons in the spinal cord. The brain begins to compensate for this on its own, working around the damage by activating alternative motor pathways that run parallel to the damaged one. These pathways may originate from the primary motor cortex on the opposite side of the brain, or from the secondary motor area immediately adjacent to the damaged area.[4]

Either way, a working connection between the brain and spinal cord can be reestablished. These new pathways are indirect, however. Under normal circumstances, the movement information sent down from the brain to the muscles passes through just one synapse—the connection between the neurons in the primary motor cortex and the motor neurons in the spinal cord. The new motor pathways involve more connections, and they activate whole groups of muscles rather than individual muscle fibers. Thus, although they can lead to an overall improvement in motor function, the recovering patient may still have difficulty moving individual fingers, for example.

Brain scanning studies show that stroke also induces long-term structural and functional changes in the primary somatosensory cortex. Chronic stroke patients exhibit increases in cortical thickness of between 4% and 13%, comparable to the volume increases seen in the mouse motor cortex after motor training and also to the structural changes associated with musical training (see chapter 6). This thickening is associated with increased cortical responses to touch sensations, and heightened sensitivity to touch in stroke patients compared to controls.[5]

Recovery from stroke involves essentially relearning how to control one's movements with these new neural pathways. The new pathways are less efficient than the damaged ones they replace, but rehabilitation can help to strengthen them and facilitate recovery of the lost functions. Intensive physiotherapy is needed in the months following the stroke, involving repetitive movements of the affected limb, which helps to strengthen the new motor pathways. Patients often lack the motivation to stick to their exercise regimes, however, and physiotherapists are currently in short supply, so in recent years rehabilitation has become increasingly reliant on robotic technology.[6]

Motor functions in stroke patients can be improved by a method called constraint-induced therapy, which involves forcing the patient to use the weakened limbs as much as possible by tying back the unaffected one.[7] But there are major individual differences in the extent to which stroke patients recover. In about one-third of patients, rehabilitation can lead to significant improvements in both movement and speech; in another third, the improvements are much smaller; and in the remainder, little or no improvement is seen.

The reasons for these varied outcomes are still not clear, but genetic and environmental factors are likely to play a role. The timing of diagnosis and treatment is also crucial—the oxygen deprivation that occurs as the result of a stroke kills millions of brain cells every minute, so quick intervention minimizes the extent of the damage, and it is now clear that the earlier rehabilitation begins, the better the outcome for the patient.

One promising approach to rehabilitating stroke patients involves altering the balance of activity between the left and right hemispheres of the brain. Normally, the hemispheres reciprocally inhibit each other by means of fibers that cross the brain in the corpus callosum, in order to coordinate the movements of all four limbs. Shortly after a stroke, the unaffected hemisphere can become more active, perhaps because of reduced cross-inhibition by the damaged side. By the same token, overactivity of the damaged hemisphere may interfere with rehabilitation.

This balance can be perturbed with TMS, which uses a magnetic coil to deliver magnetic fields to a specific part of the brain. The magnetism generates an electric field that last for about one tenth of a second, which increases or inhibits the activity of cells in the region being targeted. Studies are beginning to show that using TMS to disrupt the activity of one half of the brain can facilitate recovery, but so far the results are variable. In some patients, inhibiting activity on the unaffected side of the brain improves motor function in the affected limbs, but in others it does not.

There is, however, some evidence that the brain hemispheres can switch from inhibiting to exciting each other, at some point after the stroke, in order to facilitate recovery. Thus, using TMS to suppress activity in the damaged hemisphere, or to enhance activity in the unaffected side, can facilitate recovery by enhancing motor activity in the damaged side, but this same treatment could be counterproductive if delivered after the switch to cross-activation.[8] Learning more about how the brain adapts to a stroke on its own will undoubtedly help clinicians to improve the efficacy of such treatments.

tDCS is another noninvasive method being used to modulate brain activity. This involves using scalp electrodes to apply low-amplitude direct currents to discrete areas of the brain, and we now know that these currents induce LTP in the targeted area.[9] Both tDCS and TMS are now widely used in the clinic, as adjuncts for rehabilitation treatment and, because they can also be used to evaluate neuronal activity and brain connectivity, for diagnostic and prognostic purposes.[10]

Functional neuroimaging is increasingly being used to assess stroke damage and predict the extent to which patients might recover, too. fMRI studies show, for example, that the more a patient's movements are impaired, the more active are their secondary motor areas on the damaged side during simple gripping tasks. TMS-induced disruption of activity in this brain region impairs movement in stroke patients but not in healthy controls, suggesting that these secondary motor regions make an important contribution to recovery. Conversely, interfering with the activity of secondary motor areas on the unaffected side is far more disruptive in severely affected patients, suggesting that they are more reliant on those new pathways than are patients who suffered less damage.[11]

Some researchers are also investigating whether noninvasive brain stimulation techniques could be used to rehabilitate language functions. In most people, language functions are localized to specific regions of the left frontal and temporal lobes, and the left hemisphere is said to be the dominant one (see chapter 1). These areas are often damaged by stroke, and consequently about 20% to 40% of patients experience severe language deficits after a stroke.

Compensatory plasticity in the brain's language networks appears to be similar to that seen in the motor pathways. Damage to the language centers can lead to recruitment of surrounding areas in the damaged left hemisphere, to recruitment of dormant language centers in the right hemisphere, or both. Because language function is usually lateralized to the left hemisphere, and because loss of cross-inhibition between the hemispheres is thought to facilitate recovery, interfering with the balance of activity in the left and right hemispheres may be the key to recovery of language.

This research is still in its early stages, however, and so far the approach has produced conflicting results. As with recovery of motor function, a better understanding of how spontaneous compensatory plasticity changes with time could eventually help to optimize such treatments and improve patients' outcomes.[12]

Other research shows that early prescription of fluoxetine (Prozac) and related antidepressants enhances motor recovery after three months in stroke patients undergoing physiotherapy. It's still not clear why this is the case, however. This group of drugs is known to have anti-inflammatory effects, which may protect the patient's brain from further damage; they may also facilitate relearning by promoting LTP in newly formed motor pathways.[13]

# ADDICTION AND PAIN

The brain's capacity for neuroplasticity underlies our ability to learn from experience in order to form memories and acquire new skills, and also our ability to adapt and recover from brain injury, or at least to compensate for and work around any damage that has occurred. But the relationship between brain and behavior is not one-sided. Our experiences and behaviors induce plastic changes in the brain, and these in turn can influence our future behavior and experiences. And the consequences of neuroplasticity are not always desirable.

Addiction and pain are the best understood examples of conditions involving maladaptive forms of neuroplasticity. Addictive drugs activate and hijack the brain's reward system, and the resulting changes can remain long after the substance has been cleared from the brain, leading to cravings and to compulsive, drug-seeking behavior. Prolonged pain can induce reorganization of the spinal cord

Addictive drugs activate and hijack the brain's reward system, and these changes remain in place long after the substance has been cleared from the brain, leading to cravings and to compulsive, drug-seeking behavior.

circuitry involved in processing and then transmitting painful stimuli up to the brain, and these changes can similarly persist long after the stimuli that initially caused the pain have been removed, resulting in chronic pain states that can persist for months or years.

## Reward, Motivation, and Addiction

Addictive narcotic and prescription drugs act on and modify brain systems involved in reward and motivation. The most important of these systems is the mesolimbic pathway, which begins in a small region of the midbrain called the ventral tegmentum. In the human brain, the ventral tegmentum contains approximately 400,000 neurons. These cells synthesize and release the neurotransmitter dopamine and project their long axonal fibers to the nucleus accumbens, part of a set of subcortical structures called the basal ganglia, which are involved in procedural learning, habit formation, and the control of voluntary movement. The nucleus accumbens in turn projects to numerous other brain regions, including parts of the cerebral cortex involved in memory and decision making and the amygdala, a small, almond-shaped structure involved in fear, anxiety, and assigning emotions to our experiences.

Normally, these structures cooperate to translate motivation into goal-directed actions in order to obtain natural rewards such as food, water, and sex. The nucleus accumbens plays a central role in these processes. Everything that we find pleasurable causes ventral tegmentum neurons to fire and release dopamine into the nucleus accumbens, which then evaluates how rewarding it is, according to the amount of dopamine released. For this reason the nucleus accumbens is popularly referred to as the brain's "reward center," and dopamine as "the pleasure molecule," although they both serve numerous other functions as well.[1]

All addictive drugs target the ventral tegmentum and act on it in one way or another to enhance dopamine transmission, increasing the concentration of the neurotransmitter both there and in the nucleus accumbens and its other projection areas. Nicotine increases the firing rate of dopamine-producing ventral tegmentum neurons by acting on nicotinic receptors expressed on their surface; opioids, cannabinoids, and benzodiazepines increase their firing rate indirectly, by inhibiting the activity of GABA-producing interneurons in the ventral tegmentum; and psychostimulants such as cocaine, amphetamines, and ecstasy block the dopamine transporter, a membrane protein that normally reabsorbs dopamine once it has been released by neurons into the synaptic cleft.

Drugs hijack the reward pathway because they are more effective than natural rewards at enhancing dopamine release in the mesolimbic pathway. A single dose of

Since all pleasurable activities enhance dopamine release in the nucleus accumbens, they, too, can become addictive, and there is now evidence that activities such as gambling, sex, and shopping can lead to similar brain changes, causing people to perform them compulsively.

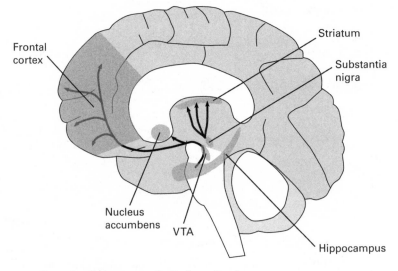

**Figure 5** The human mesolimbic (reward) pathway.

cocaine, morphine, nicotine, alcohol, or benzodiazepines induces LTP in the ventral tegmentum (see chapter 3), which persists for up to a week. Addictive substances can also produce structural changes to nerve cells, too: administration of cocaine or a related stimulant increases, whereas chronic administration of morphine decreases, the density of dendritic spines in the ventral tegmentum. Most of these findings come from experiments performed in slices of midbrain tissue dissected from the mouse brain, but brain scanning studies in humans confirm that addictive drugs increase the concentration of dopamine in the nucleus accumbens and that the increase is closely associated with the pleasurable effects of the substances.[2,3]

Addiction can be thought of as a transition from recreational use, when the drug is taken voluntarily for its pleasurable effects, to habitual use, when control over intake is lost and the user becomes compelled to find and take the drug even though it might have adverse effects. Once the drug has taken hold, the addict enters a vicious cycle, bingeing on the drug to get high, but then beginning to experience withdrawal symptoms, which trigger craving that causes the user to seek out and take more of the drug.

It is currently thought that the progression from recreational user to addict is accompanied by a sequence of functional and structural brain changes within the reward pathway. Thus, initial use of an addictive substance induces LTP in the ventral tegmentum and nucleus accumbens, producing euphoric effects. With continued use, changes begin to occur in those parts of the pathway involved in memory and executive function. The user learns to associate drug use with certain environments, people, and paraphernalia, and every subsequent dose reinforces the behaviors that lead to drug taking. The brain adapts in such a way as to make the user overvalue the rewarding effects of the drug, and the use of it becomes habitual and compulsive.[4]

Since all pleasurable activities enhance dopamine release in the nucleus accumbens, they, too, can become addictive, and there is now evidence that activities such as gambling, sex, and shopping can lead to similar brain changes, causing people to perform them compulsively. We now know that prescription medications used to treat Parkinson's disease can dramatically affect such behaviors. Parkinson's is caused by the degeneration of dopamine-producing cells in another midbrain area called the substantia nigra, leading to movement deficits and cognitive problems. Some of these symptoms can be alleviated by drugs that increase dopamine levels in the brain, but because these drugs also act on the mesolimbic pathway, they can, in rare cases, lead to pathological gambling, hypersexuality, and other compulsive behaviors.[5]

**The Pain Pathway**

Physical pain serves the evolutionarily ancient and important function of alerting us to potentially life-threatening injuries. But it too can produce long-lasting adaptations in the nervous system—changes that may contribute to various forms of persistent, pathological pain.

Our ability to perceive noxious stimuli is mediated by primary sensory neurons of the peripheral nervous system. These pain-sensing neurons have their cell bodies clustered in the dorsal root ganglia, which lie just outside the spinal cord. They have a single fiber that splits in two close to the cell body. One branch extends out to a specific patch just beneath the skin surface; it contains various receptors that are sensitive to specific kinds of painful stimuli such as excessive mechanical pressure, noxious hot and cold temperatures, and certain ingredients of the chemical cocktail that spills out of damaged cells. The other extends a much shorter distance into the back of the spinal cord, where it forms synapses with the second-order sensory neurons that project up into the brain.[6]

When these pain-sensing neurons are activated, they produce nervous impulses that travel up into the spinal cord and are transmitted to the second-order sensory neurons in the spinal cord, which relay the signals up to the somatosensory cortex. Only when these signals have been processed do we become aware of the pain, and then act to stop it in order to prevent any further damage being done.

Plastic changes can occur at the peripheral end of pain-sensing neurons beneath the skin, as well as at the synapses they form with second-order sensory neurons in the spinal cord. Activation of the protein sensors rapidly redistributes them in the nerve terminal and alters their functional properties to lower their activation threshold. This hypersensitizes the damaged tissue, so that otherwise innocuous stimuli are perceived to be painful, which aids repair by minimizing contact with the damaged tissue. It also increases the firing rate of the pain-sensing neurons, and increases the probability of neurotransmitter release from their nerve terminals in the spinal cord.

These short-term changes are usually reversible. Under some circumstances, however, there can be longer-lasting modifications to the pain system. During inflammation, growth factors released from damaged cells can trigger the synthesis and trafficking of pain receptors and their related signaling molecules in pain-sensing neurons, sensitizing the cells to painful stimuli. Trains of impulses generated by these cells can then induce LTP at synapses in the spinal cord. This amplifies the response of the secondary sensory neurons to incoming pain signals, so that repetitive, low-frequency signals produce a progressively larger output—a process called wind-up.[7,8]

Chronic or persistent pain is also associated with functional and structural changes in the primary somatosensory cortex, but different kinds of pain and injuries effect these changes in different ways. For example, cortical representation of the painful fingers expands in carpel tunnel syndrome, perhaps exacerbating the pain felt by sufferers, while the representation of affected body part shrinks in complex regional pain syndrome, possibly through disuse. Cortical reorganization occurs in several steps: within minutes of the initial injury, previously inhibited connections are "unmasked"; later on, axonal sprouting may occur within the tissue being reorganized.[9]

# LIFELONG BRAIN CHANGES

Neuroplasticity is a lifelong process. Some forms of plasticity, such as the modification of synaptic connections, take place continuously and are vital for everyday mental functions such as learning and memory (see chapter 4). On the other hand, neurogenesis is largely restricted to prenatal development, peaking at 10 to 16 weeks of fetal development, during which time the embryonic brain generates an estimated 250,000 neurons per minute. The brain continues to produce new cells after birth, but the rate of production declines rapidly within the first few years of life. The adult brain also produces new cells but, as far as we know, its ability to do so is severely limited (see chapter 5).

Likewise, synapse formation begins in the womb. At the time of birth, an estimated 2 million synapses are formed every minute, but the activity of genes involved in synapse formation peaks at around 5 years of age. Early life experiences have significant effects on developing neuronal circuits, and those effects apparently can persist throughout life. Thus, the experiences we have as young children may predispose us toward particular patterns of behavior in adulthood.

Adolescence is similarly marked by significant changes in white matter distribution, as well as by the elimination of huge numbers of synapses, which reduces the total number of synapses in the brain by some 40%. Both of these processes are vital for proper brain maturation, and both are associated with changes in behavior. Synapse formation and pruning also occur continuously throughout life, and both are now believed to be vital for normal brain function (see chapter 4).

It appears that the aging brain also undergoes stereotypical neuroplastic changes that are associated with changes in mental function; senescence, too, seems to be marked by structural and functional brain changes that contribute not only to the progressive deterioration of mental function but also to the ability to compensate for, or work around, the changes associated with age-related cognitive decline.

## Prenatal Plasticity

Proper brain development is highly dependent upon sensory stimulation and spontaneous brain activity in infancy and early childhood. The structural and functional changes induced by this activity are critical for the emergence of healthy, functional neuronal circuits. These developmental processes are among the best known and understood of all neuroplastic changes, thanks to the pioneering work performed by Hubel and Wiesel in the 1960s (see chapter 2).

In fact, neuroplasticity begins to shape neural circuits as they are being laid down in the womb. In the middle of the second trimester (20–23 weeks) the long-range connections of the sensory systems begin to form, enabling the fetus to detect environmental stimuli of various kinds, and such stimuli can fine-tune the circuitry long before birth, just as they do in early life.

Our knowledge of prenatal plasticity is still limited, however, mostly because it is far harder to investigate these processes in utero. Early studies carried out in the 1950s used electroencephalography (EEG): electrodes were placed on the mother's abdomen to record fetal brain waves during birthing. Although rather crude, this method showed that there are distinct brain wave patterns associated with fetal distress, neurological abnormalities, and brain damage sustained at birth, suggesting that brain activity during the antenatal period can sometimes predict developmental outcomes.

Today, a growing number of studies are using functional neuroimaging techniques to investigate fetal brain function. The auditory system is currently the best studied, since it is easy to deliver sound stimuli to the fetus. Hair cells in the cochlea, which convert sound waves into electrical impulses, are functional by the middle of the second trimester; by contrast, the visual system has very little sensory input at this time and does not become fully functional until after birth. Thus, auditory discrimination abilities arise in the womb—enabling newborns to recognize sounds they heard before birth, and to distinguish their mother's voice from other peoples'—and become fully mature within the first three weeks of life.[1]

This work clearly shows that cognitive processes such as attention and memory are present before birth, and researchers are now using these various neuroimaging techniques to learn more about how and when they emerge. The work is still in its infancy, but as our knowledge and technology become more sophisticated we will surely learn much more. A better understanding of these processes could provide insights into conditions such as autism, dyslexia, and possibly schizophrenia, which some now regard as a developmental disorder.

## Early Life Experiences

A landmark study published in 2004 showed that the quality of care that rat pups receive from their mother influences their behavior as adults. Rat mothers exhibit individual differences in maternal care, with some attending to and nursing their offspring more frequently than others. Pups that are repeatedly licked and groomed during the first week of life are better able to cope with stress and fearful situations in adulthood, compared to those that had little or no contact with their mothers. These differences are associated with alterations in activity of the glucocorticoid receptor gene in the hippocampus. The glucocorticoid receptor plays a critical role in the stress response, and the pups that received high levels of care from their mothers expressed it at higher levels than those who received less attention.

These effects were attributable to *epigenetic* modifications to the DNA, which alter gene expression by changing the physical structure of the chromosomal region containing the genes. Frequent licking and grooming led to epigenetic changes that opened up the chromosomal region containing the glucocorticoid receptor gene and made it more accessible to the cell's protein synthesis machinery, whereas lack of maternal care caused different epigenetic modifications that closed off the chromosome and reduced gene activity.[2]

Epigenetic mechanisms allow nature and nurture, or genes and the environment, to interact with one another, and provide a means by which acquired characteristics can be passed down generations. A key finding of this research is that the epigenetic modifications—and the behaviors associated with them—are reversible: When rat pups born to neglectful mothers are fostered by more attentive ones, the licking and grooming they receive removes the epigenetic marks that silence the glucocorticoid receptor gene, so that their stress response is comparable to that of pups that received high-quality maternal care all along. The marks could also be reversed by treating the pups with a chemical that blocks this specific type of epigenetic modification.[3]

A follow-up study by the same researchers suggests that these findings translate to humans. They performed postmortem examinations of the brains of child abuse victims who had committed suicide as adults, and compared them to the brains of suicide victims who had not been abused as children and to those who had died of other causes. They found that the hippocampi of suicide victims who were abused as children had significantly lower levels of glucocorticoid receptor messenger RNA than those of the other two groups.[4]

In the past 15 years or so, researchers have become increasingly interested in probing the relationship between brain development and socioeconomic status. We've known for many years that poorer people tend to be unhealthier, to have less access to healthcare, and to die younger that those who are better off, and the picture emerging now is that growing up in poverty has severe and persistent effects on brain development that can affect both mental and physical health in adulthood.

... growing up in poverty has severe and persistent effects on brain development that can affect both mental and physical health in adulthood.

The human brain reaches about 80% of its adult size by 2 years of age, and its growth is nearly complete by age 10. We now know, however, that extensive plastic changes continue to take place in late adolescence and beyond.

This work shows that in general, socioeconomic status is associated with variations in the makeup and function of certain brain structures. Children from poorer backgrounds have smaller gray matter volume in the hippocampus, for example, and also exhibit differences in amygdala and prefrontal cortex activity, in comparison to those who are better off. These characteristics are associated with impairments in such domains as attention, memory, and emotional regulation.[5]

The work on socioeconomic status and epigenetics is consistent with findings from numerous animal studies on the effects of environmental enrichment, and also with those of earlier research into maternal deprivation. It appears to confirm that mental stimulation and loving relationships are essential for proper brain development, and it immediately suggests multiple interventions that could break the vicious cycle of poverty, reversing or at least minimizing the consequences of childhood neglect or abuse.

Studies of children left abandoned in impoverished Romanian orphanages seem to support this idea. The deprivation left most of these children with severe cognitive impairments and learning difficulties, but those could be at least partly reversed by foster care: the earlier a child was placed into foster care, the less severe were their deficits later on.[6] But it is very difficult to test whether the epigenetic modifications associated with early life stress are also reversible in humans, and many researchers are focusing instead on what makes some people more resilient than others to the effects of stress and early life adversity.

The brain scanning studies reveal certain associations, rather than causal relationships, between poverty and brain structure and function. Socioeconomic status is a complex notion that usually incorporates an individual's level of education, income, and occupation; growing up in poverty causes chronic stress, which itself has a dramatic effect on brain development, but it often involves numerous other factors such as malnutrition, so it is currently impossible to determine exactly which components might be influencing brain development. Nevertheless, the findings are considered by some to be conclusive and are already starting to shape policies that advocate various early years interventions.

### Adolescence

The human brain reaches about 80% of its adult size by 2 years of age, and its growth is nearly complete by age 10. We now know, however, that extensive plastic changes continue to take place in late adolescence and beyond, and therefore that the brain does not reach full maturity until the mid-twenties, or perhaps even later.

The stereotypical teenager has raging hormones and heightened emotions, greatly values the approval of peers, and may take big risks to get it. Teenagers and young adults are more at risk of developing anxiety, stress, depression, and schizophrenia, and all of this is closely related to ongoing changes that occur in the brain from late childhood to young adulthood. The prefrontal cortex in particular undergoes protracted structural and functional changes throughout adolescence, and we now know that this region does not reach full maturity until the late twenties. This brain region is often said to be the seat of intellect: it subserves executive functions such as planning, decision making, and regulation of emotions.

Postmortem examination brain tissue samples from people of different ages shows that the density of dendritic spines in the prefrontal cortex increases in childhood but then begins to decrease gradually after puberty. Longitudinal studies, in which volunteers have their brains scanned repeatedly at two-year intervals, further show that gray matter density and the thickness of the prefrontal area increase in late childhood and early adolescence, peaking at age 12. Similarly, white matter volume in the prefrontal cortex increases steadily during childhood and adolescence, and then plateaus in young adulthood.

Increases and decreases in gray matter density are usually attributed to synapse formation and pruning, respectively, and changes in white matter volume to the redistribution of myelin by oligodendrocytes. Brain scans cannot confirm that any of the observed changes occur because of these processes; nevertheless, these changes slowly refine prefrontal circuits, reorganize their synaptic connections, and enhance their connectivity with other regions of the brain. Consequently, the prefrontal cortex becomes more efficient, and with that, decision making and other executive functions improve.[7]

## Parenthood

Becoming a parent is another natural experience that induces neuroplasticity. We are only just beginning to understand how parenthood changes the brain, however. Most of the work done to date has been performed in rodents, but now researchers have started using functional neuroimaging techniques to visualize how the human brain changes during pregnancy and beyond.

Newborn mice make low-frequency "wriggling" calls when hungry, and high-frequency ultrasonic vocalizations when they feel isolated. The mother learns to respond appropriately to each of these, but when she is first exposed to them, neurons in her primary auditory cortex begin to alter their responses to the sounds so that she can process and interpret them properly.

Experiments in which the activity of these cells is recorded with microelectrodes show that pups' high-frequency vocalizations increase the proportion of fast-spiking interneurons in the mother's primary auditory cortex that are tuned to their frequency. This may alter the balance of excitatory and inhibitory neurotransmission, which plays an important role in determining the extent to which the activity of auditory neuron populations is synchronized with that of cells in other brain regions.

Mere exposure to pup odors enhances the responses of neurons in the mother's primary auditory cortex to pups' vocalizations, as well as the spontaneous activity of the cells. It also seems to decrease the activity of fast-spiking interneurons, which may alter the balance between excitation and inhibition, too.

Other animal experiments show that early motherhood is associated with both structural reorganization and gray matter volume increases in various brain regions, including the prefrontal cortex, the thalamus (which relays sensory information to appropriate regions of the cortex), the hypothalamus (which synthesizes maternal hormones), the amygdala (which processes emotional information), and the striatum (which is involved in reward and motivation).

Some of these changes seem to be closely linked to the mother's attitude to her pups—bigger differences are seen in the brains of animals that interact more with their pups, compared to those that interact less. All of the changes underlie maternal behavior, priming the brain for motherhood and increasing the mother's motivation to nurture her offspring.[8,9] (In line with these findings, new mothers who suffer from postnatal depression are less sensitive to their babies' crying, and they also exhibit decreased connectivity between key brain regions, and altered glutamate neurotransmission in the prefrontal cortex, compared to those who don't.[10]) Longitudinal neuroimaging studies confirm that similar structural changes occur in human mothers shortly after giving birth. And just as a mouse mother's attitude toward her pups influences the extent of maternal brain changes, human mothers' attitudes toward their babies within the first month of birth predict the degree to which gray matter volumes increase in the following few months.

Traditionally, mothers have been the primary caregivers of their children, and fathers were considered important only in their role as breadwinners. Men, however, are becoming increasingly involved in their children's upbringing. The importance of the relationship between father and child is now recognized, and research is beginning to reveal that fatherhood too induces plastic changes in the brain.

The very first longitudinal neuroimaging study to investigate anatomical brain changes associated with fatherhood was published in 2014. This study revealed that some of the changes that take place are similar to those that occur in the brains of new mothers. The first four months after birth is an essential period for the bond between father and child to grow, and during this time increases in gray matter volume are observed in the hypothalamus, amygdala, striatum, and prefrontal cortex. By contrast, the volume of other brain regions, in the orbitofrontal cortex, cingulate gyrus, and insula, are seen to decrease. These changes are believed to be linked to changes in the father's behavior and attitude, making the attachment rewarding and strengthening the bond between father and child; precisely how these factors are related to one another, however, is still far from clear.[11]

## The Aging Brain

As we get older, most of us experience an age-related decline in mental functions such as attention, learning, memory, and task-switching, but other aspects of cognition—such as memory for facts and figures, and the ability to regulate emotions—can often improve. All of these changes can be at least partly explained by gradual changes to the structure and function of the brain.

The aging brain undergoes many such changes. In particular, neurons die off and white matter integrity decreases as we age, most evidently in the prefrontal cortex and hippocampus, which for some reason seem particularly vulnerable to the effects of aging. As a result, the volume of the prefrontal cortex decreases dramatically, and

As we get older, most of us experience an age-related decline in mental functions such as attention, learning, memory, and task-switching, but other aspects of cognition— such as memory for facts and figures, and the ability to regulate emotions—can often improve...

there is a significant reduction in the overall weight of the brain.

Research in rodents shows that aged mice have fewer synapses in the hippocampus than younger animals, and this is associated with memory impairments. Aged rats also exhibit deficits in long-term potentiation, and are more susceptible to the reversal of LTP and also to LTD. Neuronal network dynamics in the hippocampus are also compromised in aged rats, and this has been linked to deficits in spatial learning.[12]

Brain scanning studies performed on humans also reveal significant age-related differences in brain activity, but some these findings are difficult to interpret. For example, older adults sometimes exhibit greater activation of certain brain regions, even while performing just as well as younger adults on the experimental task given to them. This could mean that their brains are compensating for detrimental age-related changes by recruiting additional neural resources, but it could also reflect less efficient information processing.[13]

Thus, while we have learned a great deal about the aging brain in the past few years, it's still unclear exactly how the observed functional and structural changes are related to changes in mental function and behavior. It seems very unlikely that a single brain change is responsible for age-related cognitive decline, however, and in the future, longitudinal neuroimaging studies that repeatedly scan the same individuals as they get older will almost certainly lead to important new discoveries in this area.

There can, of course, be important differences between individuals. Most of us will eventually experience age-related cognitive decline, which is a normal part of getting older. In most people, there is also a gradual accumulation in the brain of senile plaques, one of the pathological hallmarks of Alzheimer's disease. Plaques are made up of insoluble deposits of a small protein called amyloid-beta that build up in the spaces around nerve cells. Many Alzheimer's researchers believe that plaques are toxic to brain cells, and that their deposition is the primary cause of Alzheimer's, but it is still not clear if this is the case. Plaques may be merely a consequence of the disease, rather than its cause, and some researchers now believe that it is the small fragments of amyloid-beta protein that are toxic, and that plaque formation actually protects brain tissue by sequestering these toxic fragments.

For most people, the plaque deposition that occurs with age appears to be relatively harmless, but in a small subset, brain aging seems to accelerate, leading to development of the disease. Yet others seem to be completely impervious to the effects of aging. Despite being octogenarians, these "Super Agers" outperform younger adults on memory tests, and they are found to have low densities of plaques and a thicker cortex in some brain regions, upon their death, compared to healthy controls of the same age.[14]

Such differences are likely due to a combination of genetics, environment, and life experiences. For example, Super Agers have a lower frequency of gene variants that increase the risk of developing Alzheimer's disease, and there is growing evidence that certain activities and lifestyle choices—such as physical exercise, diet, and learning a second language or musical instrument—may also offer protection against Alzheimer's and other forms of dementia.[15,16,17]

# CONCLUSION

About one hundred years ago, Santiago Ramón y Cajal, the father of modern neuroscience, stated that the adult brain is "fixed" and "immutable," and this quickly became a central dogma of the field. Cajal's own views about the brain's capacity for plasticity are, however, ambiguous, and in fact he followed this famously pessimistic statement by remarking that "it is for the science of the future to change, if possible, this harsh decree."

**Neuroplasticity as We Know It**

As we have seen in the preceding chapters, subsequent generations of neuroscientists have indeed changed the decree, by demonstrating numerous ways in which brain structure and function can change. Far from being fixed, the brain is a highly dynamic structure, which undergoes

Far from being fixed, the brain is a highly dynamic structure, which undergoes significant change, not only as it develops, but also throughout the entire lifespan.

significant change not only as it develops but also throughout the entire lifespan. *Neuroplasticity* simply means change in the nervous system, and is a collective term for all the processes that change the structure and function of the brain. Brains evolved to respond and adapt to the environment, and so neuroplasticity is an intrinsic property of nervous tissue, which occurs at all levels of organization, from the genetic to the behavioral.

The mechanisms of neuroplasticity are extremely diverse, encompassing changes in the electrical properties of neurons that last just a few milliseconds, and large-scale structural changes that develop gradually over months and years. Some modes of plasticity occur continuously, such as the strengthening, weakening, creation, and elimination of synapses; these changes are thought to be critical for learning and memory. Others are employed at specific times and places, or under special circumstances—neurogenesis occurs extensively during development but is severely restricted in adults, whereas major cortical reorganization usually occurs only as a consequence of extensive training or nerve injury. The same mechanisms can have different effects, depending on when and where they are taking place; similarly, a given effect could be brought about by different mechanisms or combinations of them.

The various types of plasticity can act separately and in concert, and each is employed in specific brain areas at certain times—and whenever else it might be needed—to ensure proper brain development, maintain normal everyday brain function, and adapt to the environment through learning and experience. As a general rule, though, the capacity for plasticity decreases with age. The brain is most malleable during development and in early childhood, when it is highly sensitive to environmental stimuli of all kinds; that malleability decreases with age, making it progressively harder to learn. This helps to explain why a 6-year-old child can go on to lead a perfectly normal life after having an entire brain hemisphere removed, but an adult cannot, and why children who learned a language (or musicians who learned to play an instrument) at a young age have more pronounced structural changes than those who learned later.[1]

Among the general public, the idea of neuroplasticity is viewed positively, and some people ascribe to it near-magical healing powers. It is because of plasticity that we are able to acquire new knowledge and skills and to recover, at least to some extent, from devastating brain injuries. Although we are beginning to learn how to enhance plasticity to facilitate recovery, this work is still in its early experimental stages, and the treatments developed thus far confer only modest benefits, if any. And neuroplasticity, although still not fully understood, must surely have limits within the physical constraints of its neurological substrates.

Neuroplasticity is not a miracle cure that will heal any ailment, transform your life, or offer infinite potential for

Among the general public, the idea of neuroplasticity is viewed positively, and some people ascribe to it near-magical healing powers ... [but] neuroplasticity is not a miracle cure that will heal any ailment, transform your life, or offer infinite potential for change.

change. It can also have negative consequences. Addiction can be thought of as a maladaptive form of learning, involving the modification of synapses within the brain's reward and motivation circuits. Likewise, synaptic modifications in the pain pathway are responsible for certain chronic pain conditions. And the prolonged period of heightened plasticity that occurs in adolescence, while vital for maturation of the prefrontal cortex, also makes teens more vulnerable to addiction and mental illness.

## Novel Forms of Neuroplasticity

The human brain is, as the cliché goes, the most complex object in the known universe. As such, it does not yield its secrets easily, so our understanding of neuroplasticity, and of brain function in general, is still very poor. Even as they struggle to understand the known forms of neuroplasticity, investigators continue to stumble upon novel mechanisms, some of which defy our long-held assumptions about how the brain works.

Take myelin, for example—the fatty tissue produced by oligodendrocytes in the brain and by Schwann cells in the peripheral nervous system. Each brain oligodendrocyte has a few extensions that form a large flat sheet of myelin, each of which wraps itself around a short segment of a single axonal fiber. Individual axons in the brain are thus insulated by numerous short segments of myelin, originating from many different oligodendrocytes and separated by the tiny lengths of bare fiber called nodes of Ranvier. This arrangement speeds up the conduction of nervous impulses along the fiber, by allowing them to jump from one node to another.[2]

Myelin is critical for the conduction of nervous impulses in the brain—as is evident from the devastation caused by multiple sclerosis and polio, both of which involve myelin degeneration. Given its importance, the distribution of myelin throughout the brain is thought to be highly stable. We have seen that neural pathways can be strengthened, and new ones created, in response to extensive training or serious injuries such as stroke; both processes involve the addition of newly formed myelin, but this happens gradually over weeks, months, or even longer periods.

A growing body of animal research now suggests that myelin redistribution can take place on much shorter timescales, however. For example, briefly training adult mice to run on a rotating wheel temporarily accelerates the production of oligodendrocytes in the brain, and blocking this new cell growth prevents the animals from mastering the new skill.[3] Other recent research shows that neurotransmitter release regulates the number of myelin sheaths formed by individual oligodendrocytes, and that oligodendrocytes preferentially wrap newly formed myelin around electrically active axons, suggesting that myelin can be redistributed in an activity-dependent manner. Short-term changes in myelin distribution could affect the extent of synchronicity between distant brain regions—a property that is increasingly regarded as an important aspect of information processing.[4,5]

Researchers continue to debate how many different types of neurons there are in the brain, and the cell types are classified in various ways, but there is general agreement that once a brain cell has matured, its identity remains fixed. Research published in the past few years, however, shows that neuronal identity can change, too. It's thought that most neurons synthesize and release just one neurochemical transmitter, and so they can be classified as "dopaminergic," "GABAergic," or "glutamatergic," according to which one they use. But it is now clear that at least some neurons can use more than one transmitter and, more surprisingly, that mature neurons can switch the transmitter they use, converting their excitatory synapses into inhibitory ones, or vice versa.[6]

Neurons can also be classified according to their electrical properties. For example, basket cells, the interneurons that control closure of the critical period in the visual cortex, are believed to exist in as many as 20 different types, the best known being the "fast-spiking" and "slow-spiking" ones, characterized according to the time frames of their responses. But it turns out that these cells can switch back and forth between fast- and slow-spiking activity, in response to neuronal activity. They appear to be constantly tuned in to neuronal network activity, and to change their firing properties in response by means of a protein that enters the nucleus and regulates the expression of potassium channels, which determine the cell's firing rate. This suggests that the 20 apparently different types basket cells are actually one and the same, and that they morph along a continuum in an activity-dependent manner. Basket cells form networks that modulate neuronal network activity, and so this identity-switching mechanism could significantly impact neuronal population dynamics by altering the ratio of fast- to slow-spiking cells within a given network of neurons.[7]

Because of this diversity of mechanisms, neuroscientists still have not fully defined neuroplasticity, and as yet there is no general theory for it. So, many questions remain. For example, are different types of plasticity somehow linked by common underlying mechanisms, so that any given experience induces a set of related changes across multiple levels of organization? Or are there situations in which a particular type of plasticity can occur independently of others? Such questions are difficult to answer, because while researchers can use microscopes to examine cellular changes in the brains of experimental animals, and neuroimaging to visualize large-scale structural changes in humans, they cannot (so far) simultaneously analyze changes at multiple organizational levels.[8]

Ultimately, neuroscientists hope to bridge the chasm between molecular events and behaviors and thought processes, and to understand how they relate to one another. The brain is increasingly viewed as one vast network containing several hundred richly interconnected "hubs," and huge amounts of money and effort are now being spent mapping brain connectivity at multiple scales. At smaller scales, brain connectivity appears to be constantly changing, but at larger scales it appears much more stable. But as we have seen, even apparently stable structures, such as long-range white matter tracts, are subject to changes that occur over longer time periods.[9]

Neuroplasticity therefore poses something of a challenge to those mapping brain connectivity, because it is still not clear exactly which types of changes are most closely correlated to our behaviors—or which scale of connectivity would be the most useful to map. Furthermore, although many similarities exist between the brains of individuals, there are also important differences. This is likely true of neuroplasticity, too: individual brains may differ in their capacity for plastic changes, so that the same experiences could induce different extents of neuroplasticity, and different types of plastic changes, in different people.

Thus, although the neuroplastic changes that occur in response to losing one's sense of sight or hearing are well documented (see chapter 1), researchers occasionally describe patients in whom they do not occur. For example, a team of psychologists in the United States recently described the case of a patient known as M.M., who had been blind between the ages of 3 and 46 years. In 2000, he underwent a corneal transplant and stem cell surgery, which restored vision in one of his eyes. Tests carried out in the two years following surgery revealed that he still had severe amblyopia, however, and a decade later his ability to recognize objects and faces remained severely impaired.[10]

In fact, the structural and functional differences between individual brains probably outweigh their similarities. It's very likely that no two brains are alike and, therefore, that there is no such thing as a "textbook brain." Your brain is, to a large extent, unique, custom-built from the life experiences you have had since being in your mother's womb, to meet the demands you place on it today. Neuroplasticity therefore lies at the heart of what makes us human, and of what makes each of us different from everyone else.

**Chapter 1**

1. Rosenzweig, M. R. 1996. Aspects of the search for neural mechanisms of memory. *Annual Review of Psychology* 47: 1–32.

2. Costandi, M. 2006. The discovery of the neuron. *Neurophilosophy* blog, 29 August, 2006. https://neurophilosophy.wordpress.com/2006/08/29/the-discovery-of-the-neuron/.

3. Rosenzweig, M. R. 1996. Aspects of the search for neural mechanisms of memory. *Annual Review of Psychology* 47: 1–32.

**Chapter 2**

1. Finger, S. 1994. *Origins of Neuroscience: A History of Explorations into Brain Function.* Oxford University Press.

2. Costandi, M. 2008. Wilder Penfield: Neural cartographer. *Neurophilosophy* blog. https://neurophilosophy.wordpress.com/2008/08/27/wilder_penfield_neural_cartographer/.

3. Bach-y-Rita, P., C. C. Collins, F. A. Saunders, B. White, and L. Scadden. 1969. Visual substitution by tactile image projection. *Nature* 221(5184): 963–964.

4. Thaler, L., S. R. Arnott, and M. A. Goodale. 2011. Neural correlates of natural human echolocation in early and late blind echolocation experts. *PLoS ONE* 6(5): e20162. DOI: 10.1371/journal.pone.0020162.

5. Striem-Amit, E., and A. Amedi. 2014. Visual cortex extrastriate body-selective area activation in congenitally blind people "seeing" by using sounds. *Current Biology* 24(6): 687–692.

6. Voss, P., and R. J. Zattore. 2012. Organization and reorganization of sensory-deprived cortex. *Current Biology* 22(5): R168–173.

7. Sadato, N. 2005. How the blind "see" braille: Lessons from functional magnetic resonance imaging. *Neuroscientist* 11(6): 577–582.

8. Lyness, R. C., I. Alvarez, M. I. Sereno, and M. MacSweeney. 2014. Microstructural differences in the thalamus and thalamic radiations in the congenitally deaf. *NeuroImage* 100: 347–357.

9. Ward, J., and T. Wright. 2014. Sensory substitution as an artificially acquired synaesthesia. *Neuroscience and Biobehavioral Reviews* 41: 26–35.

10. Zembrzyckia, A., C. G. Perez-Garcia, C.-F. Wang, S.-J. Choub, and D. D. M. O'Leary. 2014. Postmitotic regulation of sensory area patterning in the mammalian neocortex by Lhx2. *Proceedings of the National Academy of Sciences* 112(21): 6736–6741.

**Chapter 3**

1. Purves, D., and J. W. Lichtman. 1985. *Principles of Neural Development*. Sinaeur.

2. Hamburger, V., and R. Levi-Montalcini. 1949. Proliferation, differentiation and degeneration in the spinal ganglia of the chick embryo under normal and experimental conditions. *Journal of Experimental Zoology* 111(3): 457–502.

3. Cohen, S., R. Levi-Montalcini, and V. Hamburger. 1954. A nerve growth stimulating factor isolated from sarcomas 37 and 180. *Proceedings of the National Academy of Sciences USA* 40(10): 1014–1018.

4. Aloe, L. 2004. Rita Levi-Montalcini: The discovery of nerve growth factor and modern neurobiology. *Trends in Cell Biology* 14 (7): 395–399.

5. Harrington, A. W., and D. D. Ginty. 2013. Long-distance retrograde neurotrophic factor signaling in neurons. *Nature Reviews Neuroscience* 14(3): 177–187.

6. Yamaguchi, Y., and M. Miura. 2015. Programmed cell death in neurodevelopment. *Developmental Cell* 32 (4): 478–490.

7. Kandel, E. R., J. H. Schwartz, and T. M. Jessell. 1995. *Essentials of Neural Science and Behavior*. Appleton & Lange.

8. Webb, S. J., C. S. Monk, and C. A. Nelson. 2001. Mechanisms of postnatal neurobiological development: Implications for human development. *Developmental Neuropsychology* 19(2): 147–171.

9. Petanjek, Z., M. Judaš, G. Šimić, M. L. Rašin, H. B. M. Uylings, P. Rakic, and I. Kostović. 2011. Extraordinary neoteny of synaptic spines in the human prefrontal cortex. *Proceedings of the National Academy of Sciences* 108(32): 13281–13286.

10. Selemon, L. D. 2013. A role for synaptic pruning in the adolescent development of executive function. *Translational Psychiatry* 3: e238.

11. Hubel, D. H. and T. N. Wiesel 1959. Receptive fields of single neurones in the cat's striate cortex. *Journal of Physiology* 148(3): 574–591.

12. Hubel, D. H., and T. N. Wiesel. 1962. Receptive fields, binocular interaction and functional architecture in the cat's visual cortex. *Journal of Physiology* 160(1): 106–154.

13. Hubel, D. H., and T. N. Wiesel. 1965. Binocular interaction in striate cortex of kittens reared with artificial squint. *Journal of Neurophysiology* 28(6): 1041–1059.

14. Wiesel, T. N., and D. H. Hubel. 1965. Extent of recovery from the effects of visual deprivation in kittens. *Journal of Neurophysiology* 28(6): 1060–1072.

15. Sugiyama, S., A. A. Di Nardo, S. Aizawa, I. Matsuo, M. Volovitch, A. Prochiantz, and T. K. Hensch. 2008. Experience-dependent transfer of Otx2 homeoprotein into the visual cortex activates postnatal plasticity. *Cell* 134(3): 508–520.

16. Hensch, T. K. 2005. Critical period mechanisms in developing visual cortex. *Current Topics in Developmental Biology* 69: 215–237.

17. Southwell, D. G., R. C. Froemke, A. Alvarez-Buylla, M. P. Stryker, and S. P. Gandhi. 2010. Cortical plasticity induced by inhibitory neuron transplantation. *Science* 327(5969): 1145–1148.

18. Bardin, J. 2012. Unlocking the brain. *Nature* 487(7405): 24–26.

**Chapter 4**

1. Kandel, E. R., J. H. Schwartz, and T. M. Jessell. 1995. *Essentials of Neural Science and Behavior.* Appleton & Lange.

2. Sheng, M., and E. Kim. 2011. The postsynaptic organization of synapses. *Cold Spring Harbor Perspectives in Biology* 3: a005678.

3. Südhof, T. C. 2013. A molecular machine for neurotransmitter release: Synaptotagmin and beyond. *Nature Medicine* 19(10): 1227–1231.

4. Kandel, E. R., J. H. Schwartz, and T. M. Jessell. 1995. *Essentials of Neural Science and Behavior*. Appleton & Lange.

5. Sheng, M., and E. Kim. 2011. The postsynaptic organization of synapses. *Cold Spring Harbor Perspectives in Biology* 3: a005678.

6. Rosenzweig, M. R. 1996. Aspects of the search for neural mechanisms of memory. *Annual Review of Psychology* 47: 1–32.

7. Bliss, T. V., and T. Lømo. 1973. Long-lasting potentiation of synaptic transmission in the dentate area of the anaesthetized rabbit following stimulation of the perforant path. *Journal of Physiology* 232(2): 331–356.

8. Kandel, E. R., J. H. Schwartz, and T. M. Jessell. 1995. *Essentials of Neural Science and Behavior.* Appleton & Lange.

9. Malenka, R. C. 2003. The long-term potential of LTP. *Nature Reviews Neuroscience* 4(11): 923–926.

10. Malinov, R., and R. C. Malenka. 2002. AMPA receptor trafficking and synaptic plasticity. *Annual Review of Neuroscience* 25: 103–126.

11. Sheng, M., and E. Kim. 2011. The postsynaptic organization of synapses. *Cold Spring Harbor Perspectives in Biology* 3: a005678.

12. Lüscher, C., and R. C. Malenka. 2011. Drug-evoked synaptic plasticity in addiction: From molecular changes to circuit remodeling. *Neuron* 69(4): 650–663.

13. Morris, R. G., E. Anderson, G. S. Lynch, and M. Baudry. 1986. Selective impairment of learning and blockade of long-term potentiation by an *N*-methyl-D-aspartate receptor antagonist, AP5. *Nature* 319(6056): 774–776.

14. Tonegawa, S., M. Pignatelli, D. S. Roy, and T. J. Ryan. 2015. Memory engram storage and retrieval. *Current Opinion in Neurobiology* 35: 101–109.

15. Yuste, R. 2015. The discovery of dendritic spines by Cajal. *Frontiers in Neuroanatomy* 9(18). DOI: 10.3389/fnana.2015.00018.

16. Sala, C., and M. Segal. 2014. Dendritic spines: The locus of structural and synaptic plasticity. *Physiological Review* 94(1): 141–188.

17. Lamprecht, R., and J. LeDoux. 2004. Structural plasticity and memory. *Nature Reviews Neuroscience* 5(1): 45–54.

18. Cichon, J., and W. B. Gan. 2006. Branch-specific dendritic $Ca^{2+}$ spikes cause persistent synaptic plasticity. *Nature* 520(7546): 180–185.

19. Nimchinsky, E. A., B. L. Sabatini, and K. Svoboda. 2002. Structure and function of dendritic spines. *Annual Review of Physiology* 64: 313–353.

20. Allen, N. J. 2014. Synaptic plasticity: Astrocytes wrap it up. *Current Biology* 24(15): R697–699.

21. Tremblay, M.-È., B. Stevens, A. Sierra, H. Wake, A. Bessis, and A. Nimmerjahn. 2011. The role of microglia in the healthy brain. *Journal of Neuroscience* 31(45): 16064–16069.

**Chapter 5**

1. Costandi, M. 2006. The discovery of the neuron. *Neurophilosophy* blog, 29 August, 2006. https://neurophilosophy.wordpress.com/2006/08/29/the-discovery-of-the-neuron/.

2. Gross, C. G. 2012. *A Hole in the Head: More Tales in the History of Neuroscience*. MIT Press.

3. Altman, J., and G. D. Das. 1965. Autoradiographic and histological evidence of postnatal hippocampal neurogenesis in rats. *Journal of Comparative Neurology* 124(3): 319–336.

4. Kaplan, M. S. 1981. Neurogenesis in the 3-month-old rat visual cortex. *Journal of Comparative Neurology* 195(2): 323–338.

5. Costandi, M. 2012. Fantasy fix. *New Scientist* 213(2854): 38–41.

6. Ibid.

7. Nottebohm, F. 1981. A brain for all seasons: Cyclical anatomical changes in song control nuclei of the canary brain. *Science* 214(4527): 1368–1370.

8. Gould, E., and C. G. Gross. 2002. Neurogenesis in adult mammals: Some progress and problems. *Journal of Neuroscience* 22(3): 619–623.

9. Reynolds, B. A., and S. Weiss. 1992. Generation of neurons and astrocytes from isolated cells of the adult mammalian central nervous system. *Science* 255(5052): 1707–1710.

10. Costandi, M. 2012. Fantasy fix. *New Scientist* 213(2854): 38–41.

11. Braun, S. M., and S. Jessberger. 2014. Adult neurogenesis: Mechanisms and functional significance. *Development* 141(10): 1983–1986.

12. Gould, E., and C. G. Gross. 2002. Neurogenesis in adult mammals: Some progress and problems. *Journal of Neuroscience* 22(3): 619–623.

13. Eriksson, P. S., E. Perfilieva, T. Björk-Eriksson, A.-M. Alborn, C. Nordborg, D. A. Peterson, and F. H. Gage. 1998. Neurogenesis in the adult human hippocampus. *Nature Medicine* 4(11): 1313–1317.

14. Knoth, R., I. Singec, M. Ditter, G. Pattazis, P. Capetian, R. P. Meyer, V. Horvat, B. Volk, and G. Kempermann. 2010. Murine features of neurogenesis in the human hippocampus across the lifespan from 0 to 100 years. *PLoS One* 5: e8809.

15. Sanai, N., A. D. Tramontin, A. Quiñones-Hinojosa, N. M. Barbaro, N. Gupta, S. Kunwar, M. T. Lawton, M. W. McDermott, A. T. Parsa, J. Manuel-García Verdugo, M. S. Berger, and A. Alvarez-Buylla. 2004. Unique astrocyte ribbon in adult human brain contains neural stem cells but lacks chain migration. *Nature* 427(6976): 740–744.

16. Sanai, N., T. Nguyen, R. A. Ihrie, Z. Mirzadeh, H.-H. Tsai, M. Wong, N. Gupta, M. S. Berger, E. Huang, J. Manuel-García Verdugo, D. H. Rowitch, and A. Alvarez-Buylla. 2011. Corridors of migrating neurons in the human brain and their decline during infancy. *Nature* 478(7369): 382–386.

17. Spalding, K. L., O. Bergmann, K. Alkass, S. Bernard, M. Salehpour, H. B. Huttner, E. Boström, I. Westerlund, C. Vial, B. A. Buchholz, G. Possnert, D. C. Mash, H. Druid, and J. Frisén. 2013. Dynamics of hippocampal neurogenesis in adult humans. *Cell* 153(6): 1219–1227.

18. Ernst, A., K. Alkass, S. Bernard, M. Salehpour, S. Perl, J. Tisdale, H. Druid, and J. Frisén. 2014. Neurogenesis in the striatum of the adult human brain. *Cell* 156(5): 1072–1083.

19. Hanson, N. D., M. J. Owens, and C. B. Nemeroff. 2011. Depression, antidepressants, and neurogenesis: A critical reappraisal. *Neuropsychopharmacology* 36(13): 2589–2602.

20. Ernst, A. and J. Frisén. 2015. Adult neurogenesis in humans: Common and unique traits in mammals. *PLoS Biology* 13(1): e1002045.

21. Vescovi, A. L., R. Galli, and B. A. Reynolds. 2006. Brain tumor stem cells. *Nature Reviews Cancer* 6(6): 425–436.

22. Costandi, M. 2012. Fantasy fix. *New Scientist* 213(2854): 38–41.

23. Casarosa, S., Y. Bozzi, and L. Conti. 2014. Neural stem cells: Ready for therapeutic applications? *Molecular and Cellular Therapies* 2: 31. DOI: 10.1186/2052-8426-2-31.

**Chapter 6**

1. Owen, A. M., A. Hampshire, J. A. Grahn, R. Stenton, S. Dajani, A. S. Burns, R. J. Howard, and C. G. Ballard. 2010. Putting brain training to the test. *Nature* 465(7299): 775–778.

2. Max Planck Institute for Human Development and Stanford Center on Longevity. 2014. *A Consensus on the Brain Training Industry from the Scientific Community*. Accessed on 4 September, 2015, from http://longevity3.stanford .edu/blog/2014/10/15/the-consensus-on-the-brain-training-industry-from-the-scientific-community/.

3. Federal Trade Commission. 2016. Lumosity to pay $2 million to settle FTC deceptive advertising charges for its "brain training" program. Accessed on 23 February, 2016, from https://www.ftc.gov/news-events/press-releases/2016/01/ lumosity-pay-2-million-settle-ftc-deceptive-advertising-charges/.

4. Münte, T. F., E. Altenmüller, and L. Jäncke. 2002. The musician's brain as a model of neuroplasticity. *Nature Reviews Neuroscience* 3(6): 473–478.

5. Mechelli, A., J. T. Crinion, U. Noppeney, J. O'Doherty, J. Ashburner, R. S. Frackowiak, and C. J. Price. 2004. Structural plasticity in the bilingual brain. *Nature* 431(7010): 757.

6. Li, P., J. Legault, and K. A. Litcofsky. 2014. Neuroplasticity as a function of second language learning: Anatomical changes in the human brain. *Cortex* 58: 301–24.

7. Costandi, M. 2014. Am I too old to learn a new language? *The Guardian*. http://www.theguardian.com/education/2014/sep/13/am-i-too-old-to-learn -a-language/.

8. Schlaug, G., L. Jäncke, Y. Huang, J. F. Staiger, and H. Steinmetz. 1995. Increased corpus callosum size in musicians. *Neuropsychologia* 33(8): 1047–1055.

9. Elbert, T., C. Pantev, C. Wienbruch, B. Rockstroh, and E. Taub. 1995. Increased cortical representation of the fingers of the left hand in string players. *Science* 270(5234): 305–307.

10. Gaser, C., and G. Schlaug. 2003. Brain structures differ between musicians and non-musicians. *Journal of Neuroscience* 23(27): 9240–9245.

11. Bengtsson, S. L., Z. Nagy, S. Skare, L. Forsman, H. Forssberg, and F. Ullén. 2005. Extensive piano practicing has regionally specific effects on white matter development. *Nature Neuroscience* 8(9): 1148–1150.

12. Roberts, R. E., P. G. Bain, B. I. Day, and M. Husain. 2012. Individual differences in expert motor coordination associated with white matter microstructure in the cerebellum. *Cerebral Cortex* 23(10): 2282–2292.

13. Driemeyer, J., J. Boyke, C. Gaser, C. Büchel, and A. May. 2008. Changes in gray matter induced by learning—Revisited. *PLoS ONE* 3(7): e2669. DOI: 10.1371/journal.pone.0002669.

14. Scholz, J., M. C. Klein, T. E. J. Behrens, and H. Johansen-Berg. 2009. Training induces changes in white matter architecture. *Nature Neuroscience* 12(11): 1370–1371.

15. Maguire, E. A., D. G. Gadian, I. S. Johnsrude, C. D. Good, J. Ashburner, R. J. S. Frackowiak, and C. D. Frith. 2000. Navigation-related structural change in the hippocampi of taxi drivers. *Proceedings of the National Academy of Sciences* 97(8): 4398–4403.

16. Woollett, K., and E. A. Maguire. 2011. Acquiring "the Knowledge" of London's layout drives structural brain changes. *Current Biology* 21(24): 2109–2114.

17. Debarnot, U., M. Sperduti, F. Di Rienzo, and A. Guillot. 2014. Expert bodies, expert minds: How physical and mental training shape the brain. *Frontiers in Human Neuroscience* 8(280): DOI: 10.3389/fnhum.2014.00280.

18. Zatorre, R. J., R. D. Fields, and H. Johansen-Berg. 2012. Plasticity in gray and white: Neuroimaging changes in brain structure during learning. *Nature Neuroscience* 15(4): 528–536.

19. Naito, E., and S. Hirose. 2014. Efficient motor control by Neymar's brain. *Frontiers in Human Neuroscience* 8. DOI: 10.3389/fnhum.2014.00594.

**Chapter 7**

1. Buonomano, D. V., and M. M. Merzenich. 1998. Cortical plasticity: From synapses to maps. *Annual Review of Neuroscience* 21: 149–186.

2. Ramachandran, V. S., and D. Rogers-Ramachandran. 2000. Phantom limbs and neural plasticity. *Archives of Neurology* 57(3): 317–320.

3. Navarro, X., M. Vivó, and A. Valero-Cabré. 2007. Neural plasticity after peripheral nerve injury and regeneration. *Progress in Neurobiology* 82(4): 163–201.

4. Pascual-Leone, A., A. Amedi, F. Fregni, and L. B. Merabet. 2005. The plastic human brain cortex. *Annual Review of Neuroscience* 28: 377–401.

5. Schaechter, J. D., C. I. Moore, B. D. Connell, B. R. Rosen, and R. M. Dijkhuizen. 2006. Structural and functional plasticity in the somatosensory cortex of chronic stroke patients. *Brain* 129(10): 2722–2733.

6. Costandi, M. 2014. Machine recovery. *Nature* 510(7506): S8–S9.

7. Pascual-Leone, A., A. Amedi, F. Fregni, and L. B. Merabet. 2005. The plastic human brain cortex. *Annual Review of Neuroscience* 28: 377–401.

8. Ibid.

9. Rohan, J. G., K. A. Carhuatanta, S. M. McInturf, M. K. Miklasevich, and R. Jankord. 2015. Modulating hippocampal plasticity with *in vivo* brain stimulation. *Journal of Neuroscience* 35(37): 12824–12832.

10. Pilato, F., P. Profice, L. Florio, R. Di Iorio, F. Iodice, D. Marisa, and D. L. Vincenzo. 2013. Non-invasive brain stimulation techniques may improve language recovery in stroke patients modulating neural plasticity. *Journal of Neurology and Translational Neuroscience* 1: 1012.

11. Ward, N. 2011. Assessment of cortical reorganisation for hand function after stroke. *Journal of Physiology* 589(23): 5625–5632.

12. Shah, P. P., J. P. Szaflarski, J. Allendorfer, and R. H. Hamilton. 2013. Induction of neuroplasticity and recovery in post-stroke aphasia by non-invasive brain stimulation. *Frontiers in Human Neuroscience* 7. DOI: 10.3389/fnhum.2013.00888.

13. Chollet, F., J. Tardy, J.-F. Albucher, C. Thalamas, E. Berard, C. Lamy, Y. Bejot, S. Deltour, A. Jaillard, P. Niclot, B. Guillon, T. Moulin, P. Marque, J. Pariente, C. Arnaud, and I. Loubinoux, (2011). Fluoxetine for motor recovery after acute ischemic stroke (FLAME): A randomized placebo-controlled trial. *The Lancet Neurology* 10(2): 123–130.

### Chapter 8

1. Koob, G. F., and N. D. Volkow. 2010. Neurocircuitry of addiction. *Neuropsychopharmacology Reviews* 35(1): 217–238.

2. Ibid.

3. Lüscher, C., and R. C. Malenka. 2012. NMDA receptor-dependent long-term potentiation and long-term depression (LTP/LTD). *Cold Spring Harbor Perspectives in Biology* 4: a005710.

4. O'Brien, C. P. 2009. Neuroplasticity in addictive disorders. *Dialogues in Clinical Neuroscience* 11(3): 350–353.

5. Dodd, M. L., K. J. Klos, J. H. Bower, Y. E. Geda, K. A. Josephs, and J. E. Ahlskog. 2005. Pathological gambling caused by drugs used to treat Parkinson's disease. *Archives of Neurology* 62(9): 1377–1381.

6. Lumpkin, E. A., and M. J. Caterina. 2007. Mechanisms of sensory transduction in the skin. *Nature* 445(7130): 858–865.

7. Woolf, C. J., and M. W. Salter 2000. Neuronal plasticity: Increasing the gain in pain. *Science* 288(5472): 1765–1768.

8. Luo, C., T. Kuner, and R. Kuner. 2014. Synaptic plasticity in pathological pain. *Trends in Neurosciences* 37(6): 343–355.

9. Gustin, S. M., C. C. Peck, L. B. Cheney, P. M. Macey, G. M. Murray, and L. A. Henderson. 2012. Pain and plasticity: Is chronic pain always associated with somatosensory cortex activity and reorganization? *Journal of Neuroscience* 32(43): 14874–14884.

**Chapter 9**

1. Anderson, A., and M. E. Thomason. 2013. Functional plasticity before the cradle: A review of neural functional imaging in the human fetus. *Neuroscience and Biobehavioral Reviews* 37(9B): 2220–2232.

2. Sweatt, J. D. 2013. The emerging field of neuroepigenetics. *Neuron* 80(3): 624–632.

3. Weaver, I. C. G., N. Cervoni, F. A. Champagne, A. C. D'Alessio, S. Sharma, J. R. Seckl, S. Dymov, M. Szyf, and M. M. Meaney 2004. Epigenetic programming by maternal behavior. *Nature Neuroscience* 7(8): 847–854.

4. McGowan, P. O., A. Sasaki, A. C. D'Alessio, S. Dymov, B. Labonté, M. Szyf, G. Turecki, and M. J. Meaney. 2009. Epigenetic regulation of the glucocorticoid receptor in human brain associates with childhood abuse. *Nature Neuroscience* 12(3): 342–348.

5. Brito, N. H., and K. G. Noble. 2014. Socioeconomic status and structural brain development. *Frontiers in Neuroscience* 8: 276.

6. Davidson, R. J., and B. S. McEwan. 2011. Social influences on neuroplasticity: Stress and interventions to promote well-being. *Nature Neuroscience* 15(5): 689–695.

7. Blakemore, S.-J. 2012. Imaging brain development: The adolescent brain. *NeuroImage* 61(2): 397–406.

8. Elyada, Y. M., and A. Mizrahi. 2015. Becoming a mother: Circuit plasticity underlying maternal behavior. *Current Opinion in Neurobiology* 35: 49–56.

9. Kim, P., J. F. Leckman, L. C. Mayes, R. Feldman, X. Wang, and J. E. Swain. 2010. The plasticity of human maternal brain: Longitudinal changes in brain

anatomy during the early postpartum period. *Behavioral Neuroscience* 124(5): 695–700.

10. McEwan, A. M., D. T. A. Burgess, C. C. Hanstock, P. Seres, P. Khalili, S. C. Newman, G. B. Baker, N. D. Mitchell, J. Khudabux-Der, P. S. Allen, and J.-M. LeMelledo. 2012. Increased glutamate levels in the medial prefrontal cortex in patients with postpartum depression. *Neuropsychopharmacology* 37(11): 2428–2435.

11. Kim, P., P. Rigo, L. C. Mayes, R. Feldman, J. F Leckman,. and J. E. Swain. 2014. Neural plasticity in fathers of human infants. *Social Neuroscience* 9(5): 522–535.

12. Burke, S. N., and C. A. Barnes. 2006. Neural plasticity in the aging brain. *Nature Reviews Neuroscience* 7(1): 30–40.

13. Grady, C. 2012. Trends in neurocognitive aging. *Nature Reviews Neuroscience* 13(7): 491–505.

14. Rogalski, E. J., T. Gefen, J. Shi, M. Samimi, E. Bigio, S. Weintraub, C. Geula, and M. M. Mesulam. 2013. Youthful memory capacity in old brains: Anatomic and genetic clues from the Northwestern SuperAging Project. *Journal of Cognitive Neuroscience* 25(1): 29–36.

15. Abutalebi, J., M. Canini, P. A. Della Rosa, L. P. Sheung, D. W. Green, and B. S. Weekes. 2014. Bilingualism protects anterior temporal lobes integrity in aging. *Neurobiology of Aging* 35(9): 2126–2133.

16. Costandi, M. 2014. Am I too old to learn a new language? *The Guardian.* http://www.theguardian.com/education/2014/sep/13/am-i-too-old-to-learn -a-language/.

17. Wong, C., L. Chaddock-Heyman, M. W. Voss, A. Z. Burzynska, C. Basak, K. I. Erickson, R. S. Prakash, A. N. Szabo-Reed, S. M. Phillips, T. Wojcicki, E. L. Mailey, E. McAuley, and A. F. Kramer. 2015. Brain activation during dual-task processing is associated with cardiorespiratory fitness and performance in older adults. *Frontiers in Aging Neuroscience* 12(7): 154. DOI: 10.3389/ fnagi.2015.00154.

**Chapter 10**
1. Steele, C. J., J. A. Bailey, R. J. Zatoore, and V. B. Penhune. 2013. Early musical training and white matter plasticity: Evidence for a sensitive period. *Journal of Neuroscience* 33(3): 1282–1290.

2. Kandel, E. R., J. H. Schwartz, and T. M. Jessell. 1995. *Essentials of Neural Science and Behavior.* Appleton & Lange.

3. McKenzie, I. A., D. Ohayon, H. Li, J. P. de Faria, B. Emery, K. Tohyama, and W. D. Richardson. 2014. Motor skill learning requires active central myelination. *Science* 346(6207): 318–322.

4. Mensch, S., M. Baraban, R. Almeida, T. Czopka, J. Ausborn, A. El Manira, and D. A. Lyons. 2015. Synaptic vesicle release regulates myelin sheath number of individual oligodendrocytes *in vivo*. *Nature Neuroscience* 18: 628–630.

5. Wake, H., F. C. Ortiz, D. H. Woo, P. R. Lee, M. C. Angulo, and R. D. Fields. 2013. Nonsynaptic junctions on myelinating glia promote preferential myelination of electrically active axons. *Nature Communications* 4: 7844.

6. Spitzer, N. C. 2015. Neurotransmitter switching? No surprise. *Neuron* 86(5): 1131–1144.

7. Dehorter, N., G. Ciceri, G. Bartolini, L. Lim, I. del Pino, and O. Marín. 2015. Tuning of fast-spiking interneuron properties by an activity-dependent transcriptional switch. *Science* 349(6253): 1216–1220.

8. Shaw, C. A., and J. A. McEachern (eds.). 2001. *Toward a Theory of Neuroplasticity*. Psychology Press.

9. Sporns, O. 2012. *Discovering the Human Connectome*. MIT Press.

10. Huber, E., J. M. Webster, A. A. Brewer, D. I. A. MacLeod, B. A. Wandell, G. M. Boynton, A. R. Wade, and I. Fine. 2015. A lack of experience-dependent plasticity after more than a decade of recovered sight. *Psychological Science* 26(4): 393–401.

**Acetylcholine**
A *neurotransmitter* molecule released at the *neuromuscular junction* and certain brain *synapses*

**AMPA receptor**
A fast-acting non-NMDA *glutamate* receptor made up of multiple subunits

**Amygdala**
A small, almond-shaped structure in the medial temporal lobes, involved in the processing of fear and other emotions

**Astrocyte**
A star-shaped glial cell found in the brain and spinal cord, which supports neurons in various ways and regulates *neurochemical transmission*. Astrocytes are the most abundant cell type in the brain

**Axon**
One of two types of nerve fiber, along which nervous impulses are propagated to the *nerve terminal* (cf. *dendrite*)

**Cell body**
The part of the neuron from which the axon and dendrites emerge, containing the *nucleus* and apparatus for synthesizing proteins

**Central nervous system**
One of two major subdivisions of the nervous system, consisting of the brain and spinal cord (cf. *peripheral nervous system*)

**Cerebellum**
The "little brain," which plays critical roles in movement, coordination, and motor skill learning, and may also contribute to cognitive functions

**Cerebral cortex**
The prominent outer region of the brain, which is subdivided into the frontal, temporal, parietal, and occipital lobes

**Corpus callosum**
The "hard body," an enormous bundle of nerve fibers connecting the left and right hemispheres of the brain to each other

**Dendrite**
One of two types of nerve fiber, at which neurons receive chemical signals from other cells (cf. axon)

**Dendritic spine**
A tiny protruberance on a dendrite that forms the postsynaptic membrane at most excitatory synapses in the brain

**DNA**
Deoxyribonucleic acid, a double-stranded molecule inside the cell *nucleus* that carries genetic information

**Dopamine**
A *neurotransmitter*, synthesized mostly in the *midbrain*, that is involved in movement, reward and motivation, and numerous other functions

**Electroencephalography (EEG)**
A neuroimaging technique that uses scalp electrodes to measure brain waves

**Ganglion (pl. ganglia)**
A cluster of nerve cells that perform a similar function

**Glial cells**
A collective term for the various nonneuronal cell types in the nervous system, such as *astrocytes*, *microglia*, and *oligodendrocytes*, that provide nutritional and structural support for neurons and also make vital contributions to information processing

**Glutamate**
An amino acid that acts as an excitatory neurochemical transmitter through its actions on AMPA, NMDA, and kainate receptors

**Gray matter**
One of two types of nervous tissue, consisting mostly of nerve cell bodies, which has a dark appearance under the microscope (cf. *white matter*)

**Hippocampus**
A region of the medial temporal lobe critical for memory formation

**Long-term potentiation (LTP)**
A process by which synaptic connections are strengthened, widely thought to be the neural basis of learning and memory

**Mesolimbic pathway**
The brain's "reward pathway," consisting of *dopamine*-producing neurons in the *ventral tegmentum*, which project axonal fibers to the *nucleus accumbens*

**Microglia**
A type of *glial cell* that act as the brain's resident immune cells, clearing away damaged tissues and pathogens, and *pruning* unwanted *synapses*

**Midbrain**
A small but major subdivision of the brain, located in the brain stem, that controls numerous functions, such as eye movements and visual and auditory reflexes, and contains several discrete regions that synthesize *dopamine*

**Myelin**
A fatty tissue synthesized by *oligodendrocytes* that wraps itself around *axon* fibers and facilitates their propagation of *nervous impulses*

**Nerve terminal**
The end of an axon, at which *neurotransmitters* are released from *synaptic vesicles*

**Nervous impulse**
An electrical signal produced by nerve cells, consisting of a reversal of the voltage across the nerve cell membrane, initiated near the cell body and propagated along the *axon* to the *nerve terminal*

**Neurochemical transmission**
The process by which nerve cells communicate with each other, involving the release of *neurotransmitter* molecules from *synaptic vesicles* lying beneath the *presynaptic membrane*, their diffusion across the *synapse*, and their binding to receptors embedded in the *postsynaptic membrane*

**Neuromuscular junction**
The *synapse* between nerve and muscle, at which motor neurons release *acetylcholine*

**Neuron**
A type of brain cell specialized to produce *nervous impulses* and release *neurotransmitters*. The human brain contains hundreds, or perhaps thousands, of different types of neurons, most of which have three basic components: a *cell body*, a single *axon* fiber, and multiple branched *dendrites*

**Neurotransmitter**
A small chemical messenger that nerve cells use to send signals to one another. The brain produces more than one hundred such chemicals, including *acetylcholine*, *dopamine*, and *serotonin*

**NMDA receptor**
A fast-acting *glutamate* receptor made up of multiple subunits, which is critical for *long-term potentiation* (LTP)

**Nucleus**
A membrane-bound organelle containing genetic instructions stored in the form of *DNA* molecules

**Nucleus accumbens**
Part of the brain's reward system, which assigns value to stimuli according to the amount of *dopamine* they release

**Oligodendrocyte**
A type of *glial cell*, found in the brain and spinal cord, that produces *myelin* (cf. *Schwann cell*)

**Peripheral nervous system**
One of two major subdivisions of the nervous system, consisting of the *ganglia* and peripheral nerves lying outside the brain and spinal cord (cf. *central nervous system*)

**Postsynaptic membrane**
A component of the *synapse*, containing receptors for *neurotransmitter* molecules released from the *presynaptic membrane*

**Presynaptic membrane**
A component of the *synapse*, from which *neurotransmitter* molecules are released

**Pruning**
The process by which unwanted *synapses* are eliminated, performed by *microglia*

**Schwann cell**
The *myelin*-forming *glial cells* of the *peripheral nervous system*

**Serotonin**
A monoamine *neurotransmitter*, synthesized from the amino acid tryptophan, with multiple functions, including the regulation of appetite and mood

**Substantia nigra**
The "black substance," a small *midbrain* nucleus that synthesizes most of the brain's *dopamine*

**Synapse**
The minuscule junction between two nerve cells, measuring just forty billionths of a meter wide, at which neurochemical transmission takes place

**Synaptic vesicle**
A membrane-bound spherical structure found in *nerve terminals*, which stores *neurotransmitter* molecules and releases them into the synaptic cleft in response to a nervous impulse

**Transcranial magnetic stimulation**
A form of noninvasive brain stimulation that uses magnetic fields to modulate activity in a specified part of the brain

**Ventral tegmentum**
A region of the *midbrain* containing neurons that produce the neurotransmitter *dopamine*, which makes up part of the brain's reward pathway

**White matter**
One of two types of tissue in the nervous system, consisting of myelinated nerve fibers and *glial cells*, which appears white under the microscope

# FURTHER READING

Aloe, L. 2004. Rita Levi-Montalcini: The discovery of nerve growth factor and modern neurobiology. *Trends in Cell Biology* 14 (7): 395–399.

Begley, S. 2009. *The Plastic Brain*. Constable.

Costandi, M. 2013. *50 Human Brain Ideas You Really Need to Know*. Quercus.

Debarnot, U., M. Sperduti, F. Di Rienzo, and A. Guillot. 2014. Expert bodies, expert minds: How physical and mental training shape the brain. *Frontiers in Human Neuroscience* 8 (280). doi:10.3389/fnhum.2014.00280.

Gross, C. G. 2012. *A Hole in the Head: More Tales in the History of Neuroscience*. MIT Press.

Kandel, E. R., J. H. Schwartz, and T. M. Jessell. 1995. *Essentials of Neural Science and Behavior*. Appleton & Lange.

Pascual-Leone, A., A. Amedi, F. Fregni, and L. B. Merabet. 2005. The plastic human brain cortex. *Annual Review of Neuroscience* 28:377–401.

Purves, D., and J. W. Lichtman. 1985. *Principles of Neural Development*. Sinaeur.

Rosenzweig, M. R. 1996. Aspects of the search for neural mechanisms of memory. *Annual Review of Psychology* 47:1–32.

Vincent, J.-D., and P.-M. Lledo. 2014. *The Custom-Made Brain: Cerebral Plasticity, Regeneration, and Enhancement*. Columbia University Press.

Yamaguchi, Y., and M. Miura. 2015. Programmed cell death in neurodevelopment. *Developmental Cell* 32 (4): 478–490.

Yuste, R. 2015. The discovery of dendritic spines by Cajal. *Frontiers in Neuroanatomy* 9 (18). doi:10.3389/fnana.2015.00018.

# INDEX

**MOHEB COSTANDI** is a freelance science writer specializing in neuroscience. He writes the Neurophilosophy blog, hosted by *The Guardian*, and is the author of *50 Human Brain Ideas You Really Need to Know*.